Auditory cues for attention management

Dissertation

zur Erlangung des Grades eines
Doktors der Naturwissenschaften

der Mathematisch-Naturwissenschaftlichen Fakultät
und
der Medizinischen Fakultät
der Eberhard-Karls-Universität Tübingen

vorgelegt von

Christiane Glatz
aus Kirchheim unter Teck, Deutschland

März 2018

Bibliographic information published by the Deutsche Nationalbibliothek

The Deutsche Nationalbibliothek lists this publication in the Deutsche
Nationalbibliografie; detailed bibliographic data are available
on the Internet at http://dnb.d-nb.de .

ISBN 978-3-8325-4730-1
ISSN 1618-3037

Logos Verlag Berlin GmbH
Comeniushof, Gubener Str. 47,
10243 Berlin
Tel.: +49 (0)30 42 85 10 90
Fax: +49 (0)30 42 85 10 92
INTERNET: https://www.logos-verlag.de

Tag der mündlichen Prüfung: 5. Juni 2018

Dekan der Math.-Nat. Fakultät: Prof. Dr. W. Rosenstiel

Dekan der Medizinischen Fakultät: Prof. Dr. I. B. Autenried

1. Berichterstatter: Prof. Dr. A. Bartels

2. Berichterstatter: Prof. Dr. H. H. Bülthoff

Prüfungskommission: Prof. Dr. A. Bartels

Prof. Dr. H. H. Bülthoff

Prof. Dr. H. A. Mallot

Prof. Dr. B. Rolke

Ich erkläre, dass ich die zur Promotion eingereichte Arbeit mit dem Titel: "Auditory cues for attention management" selbstständig verfasst, nur die angegebenen Quellen und Hilfsmittel benutzt und wörtlich oder inhaltlich übernommene Stellen als solche gekennzeichnet habe. Ich versichere an Eides statt, dass diese Angaben wahr sind und dass ich nichts verschwiegen habe. Mit ist bekannt, dass die falsche Abgabe einer Versicherung an Eides statt mit Freiheitsstrafe bis zu drei Jahren oder mit Geldstrafe bestraft wird.

Christiane Glatz

ACKNOWLEDGEMENTS

I would like to thank everyone that contributed to the success of this dissertation.

My academic career would not have been the same without the mentoring and experiences offered to me at the Max Planck Institute for Biological Cybernetics. I am immensely grateful to Dr. Lewis L. Chuang for his indispensable support, for teaching me many skills that are not only essential as a scientist but also valuable for life outside of science, and for genuinely caring for what was in my best interest as a PhD student and for my career. I am also thankful to Professor Dr. Andreas Bartels and Professor Dr. Heinrich H. Bülthoff for being part of my advisory board, for their guidance, and for providing insightful comments on my work.

I am extremely fortunate to have had colleagues from different disciplines in AGBU that provoked thoughtful discussions, provided constructive criticism from diverse perspectives, and fostered interdisciplinary work. I am especially grateful to the members of the COMMS group, in particular Nina Flad and Dr. Menja Scheer for discussing ideas and for being supportive not only in technical but also personal aspects. I am glad that colleagues can become such great lifelong friends that make the time as a PhD student so unique.

To my mentors during my research stays abroad, Dr. Stas Krupenia and Dr. Makoto Miyakoshi. I am forever thankful for introducing me into your labs and for giving me the opportunity to learn from and cooperate with you. These experiences are priceless and without them my career would look a lot different today.

Last but not least, I would like to express my gratitude to my friends and family, especially my brother Wolfgang and my parents, who have provided the love, encouragement, and distraction when needed. Finally, Miguel has been my place to come to a rest and I am indebted to his patience, reassurance, understanding, as well as his loving care.

ABSTRACT

An exhaustible supply of mental resources necessitate that we are selective for what we attend to. Attention prioritizes what ought to be processed and what ignored, allocating valuable resources to selected information at the cost of unattended information elsewhere. For this purpose it is necessary to know the conditions that help the brain decide when attention should be paid, where to and to what information.

The question that is central to this dissertation is how auditory cues can support the management of limited attentional resources based on auditory characteristics. Auditory cues can (1) increase the overall alertness, (2) orient attention to unattended information, or (3) manage attentional resources by informing of an upcoming task-switch and, therefore, indicate when to pay attention to which task.

The first study of this dissertation investigated whether different population groups might process auditory cues differently, thus resulting in different levels of alertness (1). Study two examined more specifically whether the type of auditory cue (verbal command or auditory icon) used as in-vehicle notifications can influence the level of alertness (1). Studies three and four investigated the use of a special auditory cue characteristic, the looming intensity profile, for directing attention to regions of interest (2). Here, attention orienting to peripheral events was tested within a dual-task paradigm which required attention shifts between the two tasks (3).

Throughout the studies, I show that electroencephalography (EEG) is an indispensable tool for evaluating auditory cues and their influence on crossmodal attention. By using EEG measurements, I was able to demonstrate that auditory cues evoked the same level of alertness across different populations and that differences in behavioral responses are not due to subjective differences of cue processing (Chapter 2). More importantly, I was able to show that verbal commands and auditory cues can be functionally discriminated by the brain. While both sounds are alerting they ought to be used complementary, depending on the intended goal (Chapter 3). The studies that employed the looming sound to redirect spatial attention to an unattended visual target showed a robust benefit in response times at longer cue-target intervals (Chapter 4 and 5). The looming benefit in processing visual targets is also apparent as enhanced neural activity in the right posterior hemisphere 280*ms* after target onset. Source-estimation results suggest that a preferential activation of frontal and parietal areas, which are involved in attention orienting, give rise to this looming benefit (Chapter 5). Finally, auditory cues improved performance for unattended targets but might also benefit the central visuo-motor task by only directing attention to the periphery without moving the eyes away from the visuo-motor task. This demonstrates that auditory cues also help in managing attention by preparing for task switches such that covert attention is allocated to the respective task when this task has to be performed.

Overall this dissertation demonstrates that the careful selection of auditory cues can go a long way in supporting attention management.

CONTENTS

1   INTRODUCTION   1
   1.1   Why auditory cues are useful for attention management   1
   1.2   Auditory cue characteristics   3
   1.3   Spatial attention   11
   1.4   Thesis overview and discussion   20
   1.5   Declaration of contribution   31

2   INDIVIDUAL DIFFERENCES IN RESPONDING TO AUDITORY CUES   35
   2.1   Abstract   35
   2.2   Introduction   35
   2.3   Related work   38
   2.4   Study   41
   2.5   Discussion   49
   2.6   Conclusion and outlook   51
   2.7   Acknowledgments   52

3   BRAIN RESPONSES TO SEMANTICALLY EQUIVALENT AUDITORY CUES   53
   3.1   Abstract   53
   3.2   Introduction   53
   3.3   Related work   56
   3.4   Study methods   62
   3.5   Results   67
   3.6   Discussion   68
   3.7   Conclusion and future work   73
   3.8   Acknowledgments   75

4   TEMPORAL DYNAMICS OF AUDITORY LOOMING CUES   77
   4.1   Abstract   77
   4.2   Introduction   77
   4.3   Experiment 1: Do auditory looming sounds enhance peripheral tilt-discrimination performance across its presented duration?   81
   4.4   Experiment 2: Can the sustained performance benefit of an auditory looming sound at late CTOAs be attributed to its high intensity when the visual target appears?   83
   4.5   Discussion   84
   4.6   Methods   88

5   NEURAL CORRELATES FOR THE AUDITORY CUE BENEFIT   93
   5.1   Abstract   93
   5.2   Introduction   93
   5.3   Methods   96
   5.4   Results   102

5.5 Discussion   111
5.6 Conclusion   117

6 APPENDIX: THE PERSISTENCE OF THE AUDITORY LOOMING CUE
   BENEFIT POST CUE PRESENTATION   119
   6.1 Introduction   119
   6.2 Methods   120
   6.3 Results and discussion   121

References   125

ABBREVIATIONS

EEG	Electroencephalography
ERP	Event-Related Potential
MUA	Mass Univariate Analysis
IC	Independent Component
FP	Fronto-Parietal
CTOA	Cue-Target Onset Asynchrony
TTC	Time-To-Contact
RMSE	Root Mean Squared Error
RT	Reaction Time

1

1.1 WHY AUDITORY CUES ARE USEFUL FOR ATTENTION MANAGEMENT

Managing our attentional resources is a non-trivial task. When should one pay how much attention to which information or where to? This fundamental concern has consistently motivated the formalization of attention. Colloquially speaking, it might be true that "everyone knows what attention is" (James, 1890). For current purposes, I adopt the definition that refers to attention as conscious information processing and is, therefore, "the mechanism that turns looking into seeing" (Carrasco, 2011).

Attention is characterized by its selectivity. In other words, attention determines which information is being actively processed and which is less prioritized. This is necessary due to a limited capacity of attentional resources (Posner and Boies, 1971; Chun et al., 2011). The selective nature of attention has been described as a "spotlight" (Posner, 1980). This spotlight, or focus of attention, can be directed at objects (or features), but also to a location in space (Carrasco, 2011). As the expression itself implies, paying attention comes at the cost of allocating more attentional resources to attended than unattended locations. While signal enhancement and external noise reduction can enhance information processing at the attended location (Lu et al., 2009; Carrasco, 2011), information at unattended locations can be concurrently neglected.

Auditory cues can exert a crossmodal influence of drawing unattended information into the focus of visual attention. Audition extends the attentional range of a restricted field-of-view into a 360° surround system. This supports the detection of unseen things that would otherwise go unnoticed. Without the sound of a car approaching from behind, we would not turn our head around. This is one example that demonstrates how audition can direct visual attention to neglected locations in space. The integration of auditory information reduces the load of visually monitoring all possible locations and, thus, helps us to focus on relevant information. Besides orienting the visual system, audition has also been claimed to function as a change detector (Gaver, 1989; Ferri et al., 2015). The visual system exists "in space [...] over time" (Gaver, 1989, p.71) and is, therefore, optimal at extracting spatial but not temporal information of static objects. In contrast, the auditory system exists "in time [...] over space" (Gaver, 1989, p.71) and is sensitive to temporal aspects (Bregman, 1990). For these reasons, the auditory system complements and interacts with our visual system to provide a 'complete' representation of the environment.

Neurophysiological techniques, such as electroencephalography (EEG), have often been employed to understand how vision and audition interact in terms of brain responses. While behavioral studies can allow us to infer user behavior from the response to an event, in terms of response times (RTs) and accuracy, EEG records the electrical activity in the brain while the event itself is processed for information. The approach of evaluating event-related potentials (ERP) seeks to emphasize event-specific activity while removing unrelated activity. It achieves this through averaging over repeated trials that are time-locked to the event of interest (Luck, 2005). The ERP waveform that is obtained from this process reveals various component peak deflections that are named for their polarity and position in the waveform (e.g., P3 refers to the third positive peak deflection). Crudely speaking, the early ERP components are likely to reflect sensory processing, while later ERP components reflect higher level cognitive processing that is not modality specific (Picton et al., 1976). Hence, the EEG/ERP method can lend insights into the different processing stages that lead up to a single discrete response.

Experimental manipulations have demonstrated that these ERP components can be linked to cognitive processes. For example, we can infer how strong the cognitive processing of the event is from the ERP amplitude. Take for instance the P3 component which has been linked to target processing. Its amplitude is thought to correlate with the allocation of attentional resources. When one is required to "divide" attentional resources to perform two tasks simultaneously, fewer resources are available to process the secondary task target which can result in smaller P3 amplitudes (e.g., Wickens et al., 1983). Moreover, larger P3 amplitudes to a primary target might indicate that more attentional resources are invested in the primary task.

The central question of this dissertation is how auditory cues can support the management of limited attentional resources based on their characteristics. For this purpose, this dissertation examines how different auditory cue characteristics can heighten our overall sensitivity to detect events, also known as alerting, and influence the orienting of attention. The chapters address these aspects from experimental lab studies to realistic scenarios. First, Chapter 2 shows that different target populations do not process auditory cues differently and, therefore, result in comparable levels of alertness. Subsequently, Chapter 3 demonstrates how two distinct auditory cue types that convey the same meaning can be used complementary to capture attention for different task-goals. Together, Chapter 4 and 5 address an auditory cue characteristic with high potential for shifting spatial attention, i.e. rising-intensity (looming), in terms of temporal efficiency and the neural mechanisms that it preferentially recruits. Table 2 provides an overview of the different attentional components addressed in the subsequent chapters and the auditory cues employed to investigate them.

The following sections of this chapter will review studies that investigated auditory cue characteristics (Section 1.2) and introduce the reader to basic concepts of crossmodal visuo-spatial attention (Section 1.3). The final section of this chapter (Section 1.4) will give an overview and discussion of the experiments presented in this dissertation.

		Chapters 2-3	Chapters 4-6
Attention components	Alerting	X	
	Attention orienting		X
	Task-switching		X
Auditory cues	Verbal commands	X	
	Auditory icons	X	
	Looming		X

Table 2: Overview of topics addressed in subsequent chapters. The work presented in this dissertation investigated different concepts of attention while employing three types of auditory cues. This table displays which concept of attention is addressed in which chapter of this dissertation and which auditory cue was used in which chapter to investigate these components of attention.

1.2 AUDITORY CUE CHARACTERISTICS

In this dissertation, I address how auditory cues can help to manage attention. For this, it is necessary to understand how we attend to auditory events and which factors (i.e., physical characteristics, sound type) influence perception. This understanding can help in evaluating different auditory events for their ability to alert or direct attention.

The manipulation of physical parameters can create a completely different auditory event. Besides pitch and intensity, changes in the profile dynamics, rhythm, duration, repetitiveness, and the type of on-/offset can modify the perception of an auditory event substantially. Haas and Schmidt (1995) and Haas and Edworthy (1996) demonstrated that high pitch, short presentation rates, and high intensity sounds can increase the perceived urgency to act, which results in faster response times. Nevertheless, we do not usually analyze the pitch, rhythm, or intensity when hearing a sound but instead, try to identify its source or location intuitively (Gaver, 1989). For this reason, I will not focus on low-level auditory parameters. Instead, the following section will address the extent to which different sound types can support cognition to refer to objects, events, and concepts.

1.2.1   Auditory cue types

Auditory events can be categorized into speech and non-speech sounds. Generally speaking, non-speech can be grouped into two discriminable groups: auditory icons and earcons (e.g., Graham, 1999; Keller and Stevens, 2004; Nees and Walker, 2005). Earcons are abstract, synthetically generated, tonal sequences that have no intuitive relationship with the event that they refer to. Auditory icons are environmental sounds that are representative of an object or event and, therefore, intrinsically bear an ecological relationship.

SPEECH   Auditory events can be masked in noisy environments. Nevertheless, certain parts of unattended speech can capture attention even in noisy environments. Cherry (1953) demonstrated that meaningful messages such as our own name can be detected out of many concurrent conversations at a party. Does this suggest that it is hard to ignore speech? Does it follow that speech is a potentially good notification?

Speech notifications are easy to comprehend and do not require, or only very little, learning (Leung et al., 1997; Dingler et al., 2008). This is because speech is understood automatically after acquiring proficiency in a language (usually at a young age). Even when messages are complex, speech is still informative since it can be interpreted unambigously (Graham, 1999; Selcon et al., 1995). This leads to fewer errors compared to non-verbal sounds (Saygin et al., 2005; Graham, 1999). For these reasons, speech notifications are believed to be particularly useful when quick responses are necessary (i.e., safety-critical situations; Noyes et al., 2006).

Nonetheless, verbal notifications run the risk of being masked and/or confused with concurrent speech from neighboring conversations or radio announcements (Hakkinen and Williges, 1984). To circumvent this problem, a verbal notification can be made more discriminable from real speech by changing its pitch. Pilots have been shown to easily discriminate between real and synthesized speech (Simpson and Marchionda-Frost, 1984). They judged a voice with a pitch of 90-120Hz as especially alerting. This resulted in faster responses for synthesized speech warnings (Simpson and Williams, 1980; Simpson and Marchionda-Frost, 1984).

Another shortcoming of speech notifications is that long messages take time to communicate and might be misinterpreted if participants do not process the entire message (Simpson and Marchionda-Frost, 1984). This problem could be resolved in two ways. First, the message could be shortened by removing redundant information (Simpson, 1976) or by speeding up the message (Simpson and Marchionda-Frost, 1984; Dingler et al., 2008). The second solution to make speech messages shorter is to use single words such as 'left', 'right', 'front', or 'back' (Hunt and Kingstone, 2003; Ho and Spence, 2005, 2006). When doing so, studies that investigated the intelligibility of words suggest that longer words (i.e., words with more syllables) are preferable because they result in enhanced recognition compared to shorter words (Hirsh et al., 1954; Van Coile et al., 1997).

AUDITORY ICONS   As an alternative to speech, auditory icons have also been proposed as a means to communicate a message intuitively. Auditory icons are defined by the stereotypical meaning of objects or actions that create the sound. For this reason, they are easily understood and help in creating an internal representation of the world. For instance, the sound of a shutting car door informs of the door being closed without having to see it closing. Research has shown that sound recognition is largely influenced by everyday experience and the frequency of occurrence (Ballas and Howard, 1987; Ballas, 1993). Being more familiar with the sound affords faster recognition and reduces cognitive processing, compared to abstract sounds (Belz et al., 1999). However, recognition also depends on the current context since a different context can give rise to different interpretations (McKeown, 2005). This might be because a single auditory icon can be recognized as the involved object, action, or both. Take, for instance, the sound of water drops. This sound could signal rain if the sky is dark. If there is blue sky, it could also indicate that something is leaking. Thus one sound can give rise to more than one interpretation and result in ambiguity (Graham, 1999). Unlike in vision, confusion can also arise because sounds are not always uniquely discriminable as different objects. While we can easily differentiate a peacock from an ostrich when seeing a picture, we might have more difficulty in differentiating their calls (Fabiani et al., 1996).

Hence, the utility of auditory icons as notifications depends highly on their identifiability as well as how well they can be mapped to the event they indicate (Mynatt, 1994; Fabiani et al., 1996). Lucas (1994) demonstrated that additional information on the design of auditory notifications improved recognition accuracy but not response times. Taken together, non-speech sounds are able to convey an intended message effectively.

EARCONS   In contrast to the previously mentioned sound types, earcons are abstract musical tones generated artificially without any relation to the event that they refer to. These abstract sounds can be designed by changing physical parameters, such as the rhythm or pitch, to signal a certain level of urgency (e.g., Edworthy and Hellier, 2006). Tuning a tonal sequence to convey a certain level of urgency is an advantage of earcons. In comparison, it is not as feasible to convey different levels of urgency through speech (McKeown, 2005). Also, earcons do not resemble speech or environmental sounds and might, therefore, not be confused with real world occurrences.

In spite of these advantages of earcons, their association to events is difficult and time consuming to learn (Leung et al., 1997; Dingler et al., 2008; Salces and Vickers, 2003). The learned association for earcons might not be preserved as readily as for auditory icons and speech. When participants had to recall the association between earcons and events a week later, their performance was significantly worse than for auditory icons or speech (Leung et al., 1997). These difficulties typically result in less accuracy and slower response times with earcons than speech or iconic warnings (Leung et al., 1997; Graham, 1999; Belz et al., 1999; Ulfvengren, 2003; McKeown, 2005).

Earcon-icon hybrid: looming

A sound that not only alerts but might also effectively orient attention is an auditory looming event. This looming sound can be considered an earcon-icon hybrid because it combines ecologically valid information, intrinsic to auditory icons, with synthesized sounds. This way, beneficial characteristics of both auditory cue types are combined. More specifically, looming refers to dynamically rising intensity over time that typically signals an approaching sound source. Given that ecologically valid characteristics are interpreted intuitively (Brewer, 2000), listeners should automatically understand looming sounds as approaching objects. For this, observers naturally extract information about speed and distance from the auditory signal to perceive the approaching motion.

The optical expansion of a visual stimulus creates not only the sensation of an approaching object but also provides time-to-contact (TTC) information (Lee, 1976, 2009). This is useful in everyday life, for example, when playing sports or when estimating whether there is enough time to cross the street before an approaching car arrives. It is valuable that auditory looming can also communicate TTC, especially when visual information is unavailable. In audition, TTC is mainly specified by the increasing intensity rate that is interpreted as the object's approaching velocity (Rosenblum et al., 1987; Shaw et al., 1991; Middlebrooks and Green, 1991; Bach et al., 2008). By these means, looming sounds communicate the trajectory of an approaching sound source.

Since approaching objects could pose a potential threat, looming sounds are thought to be particularly salient. This saliency is reflected in the perception of looming sounds. Looming sounds are perceived as changing faster in intensity and, therefore, approach faster than their receding counterpart (perceived to leave; Neuhoff, 1998, 2001; Olsen and Stevens, 2010; Seifritz et al., 2002; Neuhoff, 2016). Studies that asked participants to judge the arrival of an auditory object demonstrated that participants perceived the object to be closer than it actually was. Conforming to the perception of faster approaches, participants consistently underestimated the time of arrival of looming sounds (Neuhoff et al., 2009, 2012; Riskind et al., 2014; Rosenblum et al., 1987, 1993; Schiff and Oldak, 1990; Ashmead et al., 1995). Erring on the side of caution provides more time to prepare a response to the approaching object. For this reason, it has been suggested that the anticipatory error reflects the extended margin of safety around the body (Noel et al., 2015; Ferri et al., 2015). Accordingly, approaching objects enter the margin of safety earlier than if the boundary was closer to the body and, therefore, afford more preparation time to confront or avoid the approaching object.

The looming bias to err on the safe side has been shown to be even stronger for physically or mentally vulnerable individuals (Neuhoff et al., 2012; Riskind et al., 2014; McGuire et al., 2016). Listeners that had to judge the time of arrival of the looming sounds responded sooner if they were in poor physical fitness compared to individuals that were in better physical shape (Neuhoff et al., 2012) or were suffering from anxiety (Riskind et al., 2014). This suggests that weaker individuals (subconsciously or consciously) considered themselves in need of more time to prepare for the object's arrival and, hence, in need of a larger margin of safety. This might also apply to the gender differences observed for the looming bias. Females were found to underestimate the TTC for looming sounds more than men (Schiff and Oldak, 1990; Neuhoff et al., 2009) which could be due to the fact that men are, on average, physically stronger than women (Janssen et al., 2010).

The saliency of looming sounds has also been shown to bias visual perception (Leo et al., 2011; Cecere et al., 2014; Sutherland et al., 2014; Schouten et al., 2011). When participants were asked to judge whether depth-ambiguous visual objects faced towards or away from the observer, an accompanying looming sound biased perception to the former (Schouten et al., 2011). This suggests that the spatial information, intrinsic to looming sounds, exerts a crossmodal influence on vision such that ambiguous visual stimuli are perceived as approaching.

Neurophysiological studies on humans and non-human primates also suggest that there is a preferential bias for processing looming sounds. Lu et al. (2001) as well as Maier and Ghazanfar (2007) found larger activity in the auditory cortex for looming compared to receding sounds, but also the amygdala (Maier et al., 2008) and temporo-parietal areas (Seifritz et al., 2002; Bach et al., 2008, 2015) are more activated for looming than comparable sounds. The preferential activation of areas involved in space recognition and auditory motion perception mirrors the looming sound's ability to convey spatial information.

In summary, looming sounds initiate a series of protective, physiological, cognitive, and behavioral responses that do not occur in response to other auditory objects that move in any other direction than approaching. Due to the preferential neural and behavioral response to approaching spatial information, looming sounds might be good auditory cues to direct attention to relevant locations.

1.2.2 Neural correlates for auditory cue processing

To understand which sounds are more effective for which type of attention management (i.e., alerting, orienting, task-switching), it might be profitable to understand how the brain processes theses sounds. Chapter 2 and 3 of this dissertation investigated the interchangeability of different types of auditory notifications for alerting and task-management using ERP analysis.

Auditory events that remind someone to do something (i.e., notifications or cues) can occur unexpectedly. To understand how unexpected auditory notifications are processed, we rely on an experimental paradigm widely used by cognitive neuroscientists. In what is termed the *oddball paradigm*, a standard sound is presented repeatedly (e.g., 90-80%) while a deviant target sound is presented infrequently (e.g., 10-20%). The auditory event that occurs repeatedly is believed to be coded into a representative model that predicts the next event (e.g., Schröger et al., 2015). The violation of expectations by the unexpected event can be observed in its influence on various ERP components, which I will address in the following paragraphs.

Independent of whether the sound is a standard or deviant event, the early ERP components, such as the N1 and vertex-potential (N1-P2 complex), are said to reflect physical stimulus properties, such as pitch and stimulus intensity (Crowley and Colrain, 2004; Woods, 1995; Näätänen and Picton, 1987). More precisely, an enhanced amplitude of the vertex-potential was found to reflect increasing stimulus intensity (Rapin et al., 1966). When the deviant sound violates the observer's expectations because it was louder or in a different pitch than expected, an early negative difference waveform called the mismatch negativity (MMN; Näätänen et al., 1978) can be observed. It is obtained by subtracting activity to the standard event from activity to the deviant event. Since the MMN occurs independent of whether attention is paid or not (Näätänen et al., 1978; Picton, 2011), it is thought to represent an automatic detection of physical changes in the environment.

The P2 component, typically evoked by auditory events, was found to be larger if the auditory event contained features of the deviant target (Luck, 2005). This fits previous claims that the P2 reflects object discrimination (Novak et al., 1992; García-Larrea et al., 1992), as observed in various studies. For example, Tremblay et al. (2001, 2009) found enhanced P2 amplitudes after their participants were able to discriminate events that they had previously perceived as indistinguishable. Pursuant to these findings, Shahin et al. (2003) found enhanced P2 amplitudes for musical notes compared to pure tones especially in musicians. This might reflect the increased ability of musicians to discriminate between musical notes. Considering that a larger P2 amplitude for targets in an oddball paradigm represents target identification, this classification is necessary to initiate further higher-order processing of information (e.g., P3b).

Until now, I have described ERP components that are considered to be sensitive to the physical characteristics of a stimulus. The following components are said to reflect the cognitive processing of stimuli. In particular, the late positive component P3 has been considered to contain various subcomponents such as the P3a and P3b (Picton, 1992; Comerchero and Polich, 1999; Polich, 2007). The P3a occurs earlier than the P3b (also referred to as P3 or P300) at fronto-central areas while the P3b is maximal at posterior-parietal regions (Dien et al., 2004; Picton, 2011).

The P3a is thought to reflect an automatic orienting response to rare events because its occurrence is not dependent on attending to the event (Escera et al., 2000; Polich, 2007; Squires et al., 1975; Snyder and Hillyard, 1976; Riggins and Polich, 2002; Knight, 1984). That is, rare events can evoke a P3a even when observers do not pay attention to them and even if they are task-irrelevant. Therefore, the P3a is thought to arise when a mismatch to standard events is detected, which elicits a shift of attention (Escera et al., 2000; Friedman et al., 2001). Given that the P3a is often observed directly after the MMN (Squires et al., 1975), these two ERP components can be considered as the physical difference detection (MMN) and the cognitive orienting response (P3a) to rare stimuli that are not necessarily attended.

In contrast, the P3b is associated with the active processing of rare task-relevant sounds. A P3b is evoked only if targets are attended, otherwise the P3b amplitude is reduced or even absent (Picton and Hillyard, 1974). In line with this, the P3b amplitude is enhanced if the sounds are more meaningful for the participant, for example, when they have to count or respond to deviant target sounds instead of processing them passively (Hillyard et al., 1973; Picton, 2011). To detect the rare targets in an oddball paradigm, every sound has to be compared to the template of the standard sound. Hence, the P3b has been associated with stimulus categorization (Dien et al., 2004) and context-updating that is necessary when unexpected events occur (Donchin, 1981; Donchin and Coles, 1988; Polich, 2003; Fabiani et al., 1986). The enhanced P3b amplitude for rare targets might then, for example, reflect the necessary context updating of event probabilities.

One ERP component that deserves mentioning, although research here does not address it, is the N400. It is defined by a negative amplitude around 400 ms post-stimulus onset. This component is not evoked by rare sounds in the oddball paradigm but is thought to reflect semantic processing. For this reason, it has mainly been investigated in language processing studies. In this context, N400 amplitudes were shown to be larger if the word did not relate to the previous word and smaller if it did relate semantically to the previous words (Bentin et al., 1985; Kutas and Hillyard, 1989; Kotz, 2014). Given that auditory icons also convey semantic information, the N400 has been used to investigate possible conceptual relations between words and auditory icons (Van Petten and Rhein-felder, 1995; Orgs et al., 2006). For this purpose, auditory icons preceded words or vice versa with their relationship being related or unrelated. For both types, the N400 amplitude was smaller in semantically related compared to unrelated trials, independent of which sound preceded the other.

The similarity between these two sounds is also supported by the fact that they recruit largely overlapping neural networks for processing (Cummings et al., 2006; Dick et al., 2007; Maeder et al., 2001; Adams and Janata, 2002; Dick et al., 2002) without exhibiting hemispheric asymmetries between words and auditory icons (Molfese, 1983; Hansen et al., 1983). Patients suffering from brain damage in language related areas are not only impaired in word but also in environmental sound recognition (Saygin et al., 2003). Together with the ERP findings, this suggests that both sounds rely on a common semantic processing network, which results in deficits in the verbal and non-verbal domain when impaired.

1.2.3 *Summary*

This chapter introduced the reader to the different types of auditory cues (i.e., speech, auditory icons, looming sounds) that can be employed to manage limited attentional resources. In addition, I highlighted the purpose of neuroimaging, specifically ERP analysis. Based on previous work, ERP components are thought to represent different processing steps which make an effective tool to evaluate sounds for the purposes of alerting, orienting attention, and conveying meaning, thus, task-management. This background knowledge is relevant for the studies presented in Chapter 2 and 3. These studies evaluated the use of auditory icons and verbal commands as in-vehicle notifications.

1.3 SPATIAL ATTENTION

Unlike Chapter 2 and 3, Chapter 4 – 6 address the topic of spatial attention. Namely, whether auditory cues can have an influence on orienting spatial attention.

Constantly monitoring all possible locations around us is an inefficient use of our attentional resources since most locations do not contain relevant information. To administer attentional resources efficiently, attention is selectively oriented towards a targeted location in space, also referred to as spatial attention. By attending a particular location in space, performance at this location is optimized as observed in faster response times and higher accuracy to visual targets. This has been shown for the visual modality (Posner, 1980; Müller and Rabbitt, 1989) as well as the auditory modality (Spence and Driver, 1994; Mondor and Zatorre, 1995; Schröger and Eimer, 1997). More recent work has investigated whether auditory cues can direct visuo-spatial attention crossmodally. This is not a given since the visual and auditory modality encode space in their own way. Vision creates a one-to-one mapping of external space onto the retina but is limited to the visual field-of-view. In addition, visual acuity decreases with increasing eccentricity (Rosenholtz, 2011). In contrast, audition creates a coarse 360° map of the external space by integrating interaural level and time differences of sounds (Popper and Fay, 1997). If spatial information from different modalities is integrated to form a supramodal representation of space, the complementary qualities of audition could guide the restricted vision to regions of interest (Arnott and Alain, 2011). The following sections of this chapter will focus specifically on directing spatial attention crossmodally (i.e., audio-visual).

1.3.1 Crossmodal vs. multisensory integration

To appreciate the benefits that auditory cues can have on managing visuo-spatial attention resources, it is important that crossmodal attention is not confused with multisensory integration. For this reason, this section is devoted to the differences between multisensory integration and crossmodal attention.

Auditory and visual sensory information can be bound together (Ernst and Bülthoff, 2004) if perceived as coming from the same external object. We can, for example, selectively attend to a speaker's voice but, in a noisy environment, it might be just as important to attend to the speaker's lip movements. Attending to auditory and visual information jointly reduces noise and allows us to separate target events (i.e., single voice stream) more easily from non-target background noise. Perceiving different sensory information as belonging to the same object is referred to as multisensory integration.

Multisensory integration is different from crossmodal attention (McDonald et al., 2001). To begin, their spatial and temporal constraints for binding sensory events differ. While multisensory integration only occurs if two events co-occur in time and space, crossmodal attention does not have such strict constraints. That is, multisensory integration is maximal when two stimuli are temporally aligned, allowing for deviations up to 100 ms (Meredith et al., 1987). On the other hand, crossmodal attention, where one stimulus (cue) precedes another (target), operates differently. Unlike multisensory integration, cuing benefits manifest with more time between cue and target due to preparatory effects (Näätänen et al., 1974; Lu et al., 2009). In addition, crossmodal cuing effects appear even when the cue does not indicate the exact, but only approximate, target location (Stein et al., 1996; Lee and Spence, 2015). Furthermore, multisensory integration and crossmodal attention rely on separate neural regions. While multisensory integration has been observed to activate the superior colliculus (Meredith et al., 1987), crossmodal spatial attention primarily involves frontal and parietal regions (Macaluso, 2010; Petersen and Posner, 2012).

Nevertheless, crossmodal attention and multisensory integration might interact. Attention can, for example, widen the temporal window for multisensory integration such that two sensory stimuli do not need to be exactly aligned (Donohue et al., 2015). Also, multisensory integration can influence attention depending on whether it occurs at pre- or post-attentive levels (Koelewijn, 2009). Although these two processes might influence each other, the current work focuses on crossmodal spatial attention.

1.3.2 *Directing covert visual spatial attention crossmodally*

Much research has focused on orienting spatial attention within the visual modality. This research has shown that spatial attention can be allocated overtly, by moving the eyes to focus on a location, or covertly, by attending to a location in space without moving the eyes. In this dissertation I focus on the covert orienting of attention that does not involve eye movements.

Traditional spatial attention studies employ the *Posner paradigm* (Posner, 1980). In these experiments, a cue that precedes the target directs attention to a location in space. If the cue indicates the target's location (valid), response times are faster than if the cue indicates a location that is different from the target (invalid; Posner, 1980). Depending on the type of cue employed, two different attention systems might be recruited. If a salient cue is presented at the peripheral location where the subsequent target will appear, attention is captured involuntarily. This is termed exogenous. When the cue itself is positioned neutrally in space but symbolically indicates the subsequent target location, attention is oriented voluntarily because of the spatial expectation through the cue. This is termed endogenous. Voluntary attention is also referred to as sustained attention because it takes about $300ms$ to deploy but can be sustained at the endogenously cued location as long as necessary. In contrast, the capture of involuntary attention is also referred to as transient attention because it peaks early ($\approx 120ms$) and declines subsequently. At long cue-target intervals, the benefit of exogenously cued attention reverses such that response times might be slower for targets at the previously cued location than other locations. This effect is known as inhibition of return (IOR; Posner and Cohen, 1984; Posner et al., 1985; Klein, 2000). Posner and Cohen (1984) hypothesized that an exogenous cue draws attention to the indicated location and afterwards moves to another location with a bias against returning to the attended location to optimize visual search. While this phenomenon was first shown for the visual modality, it has also been shown to occur for crossmodal capture of attention (Spence and Driver, 1998; Spence et al., 1998; Tassinari and Campara, 1996).

CROSSMODAL LINKS IN ENDOGENOUSLY CUED SPATIAL ATTENTION   The type of cue, exogenous or endogenous, might influence whether attention is directed crossmodally to a location in space or not. Given that endogenous cues create an internal spatial expectancy, this expectancy might overarch different modalities. In other words, a supramodal system might govern the endogenous directing of attention. Spence and Driver (1996) induced such a spatial expectancy within one modality (primary modality) and investigated whether the cuing effects typically found for this modality would also extend to other modalities (secondary modality). As expected, their results showed a target advantage for the primary modality (e.g., vision), which was due to directing attention voluntarily to the expected location. However, they also found an advantage for targets in the secondary modality (e.g., audition). This suggests that inducing a strong spatial expectancy in one modality also shifts spatial attention to this location in other modalities. Although the effects in the secondary modality were smaller than in the primary modality, these findings put forth the idea of a supramodal system for sustaining spatial attention voluntarily.

CROSSMODAL LINKS IN EXOGENOUSLY CUED SPATIAL ATTENTION   Since exogenous cues do not evoke top-down spatial expectations (c.f., endogenous cues), the direction of spatial attention by exogenous cues might not be governed by a supramodal system. Thus, can salient external events also capture spatial attention crossmodally? Several studies have addressed the question of whether involuntary spatial attention is governed by a modality-specific or supramodal system (e.g., Schmitt et al., 2000; Spence and Driver, 1997; Ward, 1994; McDonald et al., 2000).

Early findings showing crossmodal exogenous cuing effects could be accounted for by eye-movements (e.g., Ward, 1994) or response facilitation (e.g., Simon and Craft, 1970). For these reasons, Spence and Driver (1997) conducted a study that controlled for these two possible confounds. To investigate whether spatial attention can be captured crossmodally, their spatial cues did not convey information about the correct response but only indicated the target location. Their results demonstrated that crossmodal links also exist in involuntary attention. Exogenous auditory cues were able to improve target performance at the cued location not only for auditory but also visual targets. After reviewing the existing literature on crossmodally cued spatial attention, Spence (2010) concluded that not only endogenous but also exogenous shifts of spatial attention are governed by a supramodal spatial attention system. He claimed that the occasional failures to show crossmodal links for exogenously cued attention are only artifacts due to specific experimental paradigms.

1.3.3 Neural correlates of crossmodal spatial attention

In Chapter 5 I demonstrate how a network of brain regions underlies task management in a dual-task paradigm. This task management involves the crossmodal shifting of spatial attention from the center to a peripheral location. The benefit of cuing this shift of attention can be observed in the ERP response to the stimulus presented in the periphery (see Chapter 5.4.2). Here, I will give an overview of the existing literature on ERP components for crossmodal spatial attention shifting as well as link these to previously proposed neural networks involved in spatial orienting.

To begin, ERPs have been used to study the influence of crossmodal attention on target processing. The effect of captured spatial attention is reflected in an ERP component called negative difference (Nd). It is obtained by subtracting the ERP activity for invalid cued targets from the ERP activity of valid cued targets (McDonald et al., 2000). The larger negativity for valid compared to invalid cued targets has been shown within the visual modality (Mangun and Hillyard, 1991; Eimer, 1993, 1994), within the auditory modality (Schröger, 1993, 1994; Schröger and Eimer, 1997), but also across modalities in which auditory cues were able to capture visuo-spatial attention (Eimer and Schröger, 1998; McDonald et al., 2000; Hillyard and Münte, 1984; Teder-Sälejärvi et al., 1999). Given that ERP effects of shifting spatial attention were similar across modalities, it confirms the previous proposition that a supramodal space representation mediates shifts of spatial attention (Eimer and Driver, 2001; Green and McDonald, 2006).

To investigate the process of orienting covert spatial attention itself, several studies have examined the ERP activity present in the cue-target interval. These studies revealed that endogenous cues elicit two ERP components: (1) Anterior Direction Attention Negativity (ADAN), and (2) Late Direction Attention Positivity (LDAP) at contralateral posterior electrodes (Harter et al., 1989; Mangun, 1994; Nobre et al., 2000b; Green et al., 2005). The ADAN has been associated with the visual, rather than the supramodal, control processes of spatial attention because audio-spatial attention shifts did not generate ADAN (Eimer et al., 2003). However, auditory cues evoked ADAN if the target was in the visual modality. This suggests that ADAN might only be generated if visuo-spatial attention is involved (Green et al., 2005). In contrast, the LDAP component was replicated consistently for visual, auditory, as well as crossmodal spatial attention shifts. The LDAP is, therefore, considered to be independent of modality and reflect supramodal processes for shifting attention voluntarily in space (Green et al., 2005; Green and McDonald, 2006; Eimer and Driver, 2001; Eimer et al., 2002). Together, the frontal ADAN and the parietal LDAP could reflect control mechanisms for shifting visuo-spatial attention crossmodally.

Indeed, these ERP components coincide with the fronto-parietal (FP) network for orienting attention proposed by Posner and Petersen (1990; 2012). In this network they considered parietal areas to play a key role in spatial attention. Patients with parietal lesions demonstrated a reduced ability to shift attention to cued targets, especially if they were cued at an invalid location (Petersen et al., 1989; Posner and Cohen, 1984). Furthermore, research with healthy individuals consistently confirmed the involvement of parietal areas in spatial attention (Farah et al., 1989; Gitelman et al., 1999; Nobre et al., 2000a; Davidson and Marrocco, 2000). Orienting attention to cued targets increased activity in parietal regions. This might be because an internal space representation, situated in the parietal lobe, enables the shifting of attention to specific locations in space (Mesulam, 1999). Indeed, converging evidence from neuroimaging studies suggests that the shifting of spatial attention is coordinated based on a supramodal space representation in the parietal lobe (Farah et al., 1989; Macaluso, 2010; Yang and Mayer, 2014).

The frontal areas that are involved in spatial attention shifting are thought to function as executive control (Petersen and Posner, 2012). That is, the activation of frontal regions in the FP network for spatial attention regulate the switching of attention to another location to initiate a new task (Dosenbach et al., 2008). Hence, frontal executive control is necessary to coordinate multitasking, or rather task switching.

1.3.4 *Different simultaneous tasks compete for attention*

Attention management is especially important when two concurrent tasks compete for resources. Often, these tasks are separated by different locations. For this reason, experiments in Chapter 4, 5, and 6 investigated spatial attention orienting within a dual-task paradigm. To be able to appreciate the findings of these experimental paradigms, this section will introduce prominent theories that account for how attentional resources are managed to support dual-task performance.

Our daily life often requires us to do more than one task at the same time. While it is possible to read a book and, in parallel, listen to music, it is more difficult to drink coffee and type an email at the same time. The difficulty in performing certain tasks simultaneously arises from our limited capacities for task-relevant resources. Various attention theories have tried to explain the attentional limitations that restrict our ability to do more than one thing at the same time.

The model by Broadbent (1957) suggests that all sensory information is first perceived in parallel but only some information is selected for higher cognitive processing based on stimulus properties. This results in a sequential processing of information on a higher cognitive level (Broadbent, 1957, 1971; Welford, 1952, 1967). Nevertheless, all early selection theories do not account for phenomena such as being able to read a book and listen to music at the same time. According to this approach, one would have to alternate between reading and listening which, according to our personal experience, we do not do.

Other models focus on explaining the extent to which two simultaneous tasks interfere in terms of their competition for similar/dissimilar attentional resources. Such models are based on the assumption that information processing consumes attentional resources, which reduce total available capacity. While some assume that there is only one attentional resource that is shared across all tasks (Kahneman, 1973; Moray, 1967), others advocate the idea that there are several distinct pools of attentional resources that, together, make up the overall capacity (Wickens, 1976; Kantowitz and Knight, 1976; Navon and Gopher, 1979; Wickens, 1980, 2002). These models are also referred to as multiple resource models and differ in that two concurrent tasks need not necessarily compete for the same attentional resources. Indeed, experimental evidence has shown that the competition for attentional resources depends on the similarity between concurrent tasks. If, for example, both tasks require a motor output they will interfere more than if one of the tasks requires a vocal response (e.g., Wickens, 1976; McLeod, 1977). For this reason, the difficulty in drinking coffee and typing an email in parallel might be caused by the competition for manual resources. In contrary, listening to music and reading a book target separate auditory and visual resources and, hence, do not compete. Multiple resource theories are better suited in accounting for a wider range of ecological activities. However, it raises a fundamental question: When and how do tasks conflict for their demands for which resources?

Switching tasks to perform dual-tasks

Attention management is not only determining where to pay attention to while performing two tasks but also when attentional resources need to be allocated to a specific location. Capacity limitations may prevent the parallel processing of information and, therefore, lead to serial processing. In the case of writing an email and drinking coffee, the writer stops to write in order to pick up the cup of coffee, take a sip, and then puts the cup back down before he returns to typing. This kind of dual-tasking can also be considered as task switching. The switching between tasks requires the redirection of attention from one task to another one. To do so, we need to disengage from the primary task to attend to a secondary task and back again to the primary task. Research has shown that task switching comes at a cost. This "switch cost" is manifested in either slower responses and/or more errors to the task that one switches to (Pashler, 2000; Monsell, 2003; Kiesel et al., 2010). One reason for this cost could be the necessary task set reconfiguration (Monsell, 1996) that defines the goal of the task. Initiating this reconfiguration in advance helps in preparing to perform a task (Fagot, 1994). This suggests that switch costs can be effectively mitigated by using a cue to indicate an upcoming task which allows task-set reconfiguration prior to the task switch.

Manual control task instead of fixation cross

In the context of dual-tasking it is arguable whether fixating a central cross can be considered a task. The fixation might not motivate a task switch and/or shift of attention to another task and back again. That is, covert attention does not need to be deployed to the center while fixating it. In contrast, participants need to pay covert attention when performing a challenging cognitive task in the center. The shift of attention while switching tasks is associated with a cost for central task performance.

Previous dual-task experiments have used a rapid serial visual presentation task to maintain focused attention in the center. These tasks require the semantic processing of letters or numbers that change quickly. Each stimulus appearance can function as an exogenous cue to reorient attention back to the central task (Spence, 2010). In contrast, a visuo-motor tracking task requires participants to track a random-appearing continuous signal. That is, by knowing the desired location and visually perceiving the actual location of a cursor, the discrepancy can be transformed into the required manual input to correct for the deviance. Hence, the visuo-motor tracking task covers a range of cognitive functions and is a suitable continuous task to occupy attention on various levels without acting as an exogenous cue to repeatedly redirect attention to the center.

1.3.5 Summary

The current section showed that auditory cues can orient visuo-spatial attention crossmodally. For both, voluntary and involuntary attention, this orienting is thought to be coordinated via a supramodal space representation. In the FP network for orienting attention, parietal brain areas seem responsible for the spatial component of orienting attention. In turn, frontal brain areas are thought to take on executive functions such as controlling the switching of attention between different locations and tasks. This is especially important when two tasks have to be performed concurrently. Chapters 4-6 address the use of auditory cues for redirecting visuo-spatial attention from a continuous central task to occasional peripheral visual targets. In this context, auditory cues might not only be helpful in orienting attention to a location of interest but also in determining when to switch attention from one to another task.

To recap, this dissertation investigated how auditory cues can benefit attention management such that attention is payed to locations and information when necessary. Imagine driving a car in the city center where a lot of advertisements are flashing on bill-boards. To avoid these distractions, we have to focus our attention on the street in front of us. This selective processing of our surrounding might result in the neglect of information outside the current focus. Auditory cues can heightening our overall sensitivity to detect events and direct attention to unnoticed events such as a nearing pedestrian.

In this final section, I discuss the current findings according to the contributions to the three different aspects of attention management: alerting, orienting of attention, and task-management. For this purpose, I will first discuss how different types of auditory cues can increase alertness. To continue, I will now address the question whether specific auditory cue characteristics (i.e., looming) can lead to preferential orienting of attention. Finally, I will focus on how auditory cues are useful in managing attention deployment to different tasks. Besides addressing these aspects of attention management, this section is organized in terms of questions that were answered in this dissertation.

1.4.1 Auditory cues for alerting (Chapter 2-3)

Auditory cues alert listeners by mobilizing attentional resources subsequent to spatially-unspecific cues. In the context of in-vehicle notifications, verbal commands and auditory icons have been evaluated for their effectiveness in alerting drivers by comparing response times and response accuracy to these auditory cues. Previous findings of these studies have been mixed. While some prefer verbal commands for their faster RTs (Lucas, 1994; Ho and Spence, 2005; Fagerlönn et al., 2015), others demonstrated faster RTs for auditory icons (Graham, 1999; Saygin et al., 2005). In addition, there are also studies that have found no difference between the two cue types (McKeown, 2005; Ulfvengren, 2003).

CAN VERBAL COMMANDS AND AUDITORY ICONS BE USED INTERCHANGEABLY AS NOTIFICATIONS? In Chapter 3, I used EEG measures to evaluate verbal commands and auditory icons as in-vehicle notifications based on the neural processing instead of behavioral performance only. In line with previous work that investigated the processing of auditory events, both my cues, auditory icons and verbal commands, elicited the typical N1-P2-N2-P3 ERP waveform (Crowley and Colrain, 2004; Kraus and Nicol, 2009; Picton, 2011, see also Chapter 3, Figure 5 and Figure 6). While little work has compared ERPs to words and auditory icons for the use as cues, there are studies that focused on their semantic processing (i.e., ERP component N400). For example, Van Petten and Rheinfelder (1995) had semantically related or unrelated words precede auditory icons and vice versa to show that the two share a conceptual relationship. Although they focused on analyzing the N400, their early ERP components for auditory icons and verbal commands are in congruence with our findings. Both waveforms had more negative N1 amplitudes for auditory icons than words and more positive P2 amplitudes for words than auditory icons (compare Figure 5 and 6 of Chapter 3 with Figure 1 from Van Petten and Rheinfelder, 1995). Given that a larger P2 amplitude indicates better discriminability (García-Larrea et al., 1992; Novak et al., 1992), their data supports our finding of verbal commands being discriminated better than auditory icons against background noise.

Due to the examination of the N400 component, previous studies that compared auditory icons and verbal commands did not allow for the elicitation of P3b. Contrary to these experimental designs, Chapter 3 employed the oddball paradigm which did give rise to P3b. The results demonstrated that the P3b amplitude increased for auditory icons. This suggests that auditory icons update contextual working memory less effortful than verbal commands. Together with the fact that verbal commands are discriminated better, the results show that auditory icons and verbal commands can be functionally discriminated by the brain. That is, while auditory icons and verbal commands are both alerting they might serve different functions and should not be used interchangeably. The analysis of the neurophysiological data suggests that verbal commands might be especially useful in stressful or urgent situations where notifications need to be discriminated easily. Edman (1982) and Graham (1999) previously proposed this application based on the assumption that speech is familiar and processed automatically. Also, our proposition to use auditory icons to inform of changing driving conditions, due to their less effortful updating of contextual working memory, is in line with previous recommendations. Adcock and Barrass (2004) and Kazem et al. (2003) endorsed the use of auditory icons to enhance situational awareness by using them to notify of less urgent environmental circumstances (Edworthy and Hards, 1999; Keller and Stevens, 2004).

DO DIFFERENT POPULATION GROUPS IN DIFFERENT EXPERIMENTAL SETTINGS NECESSARILY RESULT IN DIFFERENT AUDITORY CUE PROCESSING? To answer this question, I evaluated the verbal and iconic auditory notifications on naïve university students under controlled experimental settings as well as on professional truck drivers in a high fidelity vehicle simulator (see Chapter 2). Since these auditory notifications were designed for in-vehicle task-management, is it plausible that truck drivers attribute more relevance to these notifications than students? Does this relevance result in different processing of auditory cues? Although these notifications had no real-life meaning for the students, I did not find support for this supposition. Indeed, I found comparable brain activity with regards to the processing of auditory notifications, in spite of differences in age, professional expertise, and experimental conditions. This suggests that well-controlled experimental work is transferable to more realistic settings and, more importantly, that auditory cues have a robust effect on the cognitive processing.

In summary, while it has been previously shown that verbal commands and auditory icons are semantically equivalent (Van Petten and Rheinfelder, 1995; Cummings et al., 2006), the current results demonstrated that both auditory cues can alert users and evoke equally fast response times. This was independent of the experimental settings (controlled laboratory vs. realistic driving simulator). However, they ought to be employed differently depending on the intended goal. For this reason, the question whether to use verbal commands or auditory icons to alert drivers and capture their attention can be replaced by questioning what the intention of the cue is and use both types accordingly to alert drivers.

1.4.2   *Auditory cues for orienting attention (Chapter 4-6)*

To be able to process relevant information from our surrounding, we need to attend the locations where this information is presented. A considerable amount of research has investigated spatial attention and the crossmodal orienting of attention to regions of interest (see Section 1.3). These studies typically employ paradigms where the target location is cued validly (at target location) or invalidly (different than target location). Their results have repeatedly shown a valid cue benefit and an invalid cue cost. Leo et al. (2011) also used this paradigm to demonstrate that looming sounds orient visuo-spatial attention more efficiently than comparable sounds. They found higher accuracies for visual target identification when the visual target was accompanied with a looming sound. Presenting the looming sound together with the visual target does, however, not provide insights in the looming sound's functionality as a cue. Thus, the question remains whether looming sounds also enhance the direction of visuo-spatial attention when preceding the visual target.

Given that spatial attention has repeatedly been shown to be oriented to validly cued locations with greater success, the experimental design in Chapter 4, 5, and 6 only cued at target locations. Instead of investigating the benefit of cuing targets at the correct location, I investigated the benefit of looming cues on orienting attention to validly cued visual targets.

CAN LOOMING CUES SUSTAIN INVOLUNTARY SPATIAL ATTENTION?   Contrary to previous work that used looming sounds as cues (Bach et al., 2009; Burton, 2011; Ho et al., 2013; Gray, 2011), I employed looming sounds to direct spatial attention to locations of interest. Typically, exogenous cues have their largest benefit at short CTOAs (see Müller and Rabbitt, 1989). In agreement, the findings of Chapter 4 show that all my auditory cues were able to orient attention at short cue-target intervals (i.e., 250 ms CTOA). More importantly, the auditory looming cue was also able to orient attention even at later time points (i.e., 500 ms CTOA, see Chapter 4 and 5). This orienting of attention is, however, transient and not sustained voluntarily. Introducing a 500 ms gap after the sound offset such that the visual target is presented at 1000 ms CTOA eliminated the looming cue benefit (see Chapter 6). If the looming benefit at 500 ms CTOAs would reflect the voluntary orienting of attention, introducing a gap between cue and target should not have effected the looming cue benefit.

Given that only the looming cue was able to improve performance at 250 ms CTOA as well as 500 ms CTOA (see Chapter 4), it is suggestive that transient attention captured by looming sounds can be sustained. The remaining question is for how long transient attention might be sustained. To recap, I did not find faster responses to targets presented 500 ms after the auditory cue at 1000 ms CTOA. Nevertheless, it might still be possible to observe a looming cue benefit at 1000 ms CTOA if the auditory cue is 1000 ms instead of 500 ms long. Burton (2011) was able to observe a looming cue benefit at 1000 ms CTOA with a sound that was 1000 ms long. Together with my results, this suggests that the strongest looming benefit can be observed at the sound offset. As long as the object is relevant because it is approaching, attention is sustained for the duration of the looming sound. The offset of this looming sound might signal that the object has reached the observer. If there is no demand for attention at the cued location at the sound offset (i.e., no target), attention is withdrawn from the cued location. Especially when involved in another task, it is costly to sustain attention. Thus, it is reasonable to withdraw attention from the cued location and focus on the other task if the cued location becomes irrelevant. Future research should investigate more specifically whether auditory looming cues can sustain transient spatial attention until the sound offset due to the TTC information conveyed by the looming cue.

DOES OVERALL INTENSITY INFLUENCE THE AUDITORY LOOMING CUE BEN-EFIT? The preferential orienting of attention at the sound offset of a looming cue might not be surprising given that it is the point of highest intensity. Thus, auditory looming cues cannot be discussed without taking sound intensity into consideration. Most studies employing auditory looming stimuli, including the experiments featured in this dissertation, rely on the increasing intensity profile to convey the approaching auditory motion (e.g., Shaw et al., 1991; Middlebrooks and Green, 1991; Bach et al., 2008, Chapters 4-6). It has recently been confirmed that the looming bias observed when using rising-intensity stimuli actually reflects a bias towards approaching motion and is not due to physical changes in intensity (Baumgartner et al., 2017). Nevertheless, the overall intensity has been shown to influence the perception of intensity change (Olsen and Stevens, 2010; Olsen, 2014; Teghtsoonian et al., 2005). That is, loud looming sounds were judged as rising faster most likely due to higher end-intensities (Susini et al., 2002, 2010). In contrast to these findings, my results suggest that the overall intensity of auditory looming cues did not show an end-intensity bias in processing visual targets.

Unlike static cues, overall intensity does not influence the cuing effect of looming sounds. I only observed an effect of overall intensity in the first experimental session of Chapter 5 which showed to disappear with adaptation. In the second session of Chapter 5's experiment as well as in Experiment 2 of Chapter 4, I did not find an influence of intensity on the looming cue. This was also reflected in the ERP amplitude of the auditory cue benefit (at 280 ms). The amplitude for looming sounds was larger than for static sounds, independent of the cues' overall intensity (Chapter 5, Figure 17). Nevertheless, static cues were able to decrease response times with increasing intensity (Chapter 4, Experiment 2; Chapter 5, Session 1). This effect of intensity is unsurprising but might also reverse into startling listeners (e.g., Ludewig et al., 2003). Nonetheless, I was able to show that static cues are also prone to intensity adaptation. These factors render the use of high intensity levels for improved performance impractical. Throughout this dissertation, the looming sound proofed to be a more robust auditory cue than the static sound because neither intensity differences nor training effects influenced the looming cue benefit.

WHAT ARE THE NEURAL CORRELATES FOR THE AUDITORY LOOMING CUE BENEFIT IN SHIFTING VISUO-SPATIAL ATTENTION? Previous research has investigated the multisensory integration of looming sounds. These findings have shown repeatedly that looming sounds bias visual processing (Romei et al., 2009; Leo et al., 2011; Schouten et al., 2011; Sutherland et al., 2014). In addition, neurophysiological studies have identified neural regions that are enhanced by the multisensory integration of the looming property (Cappe et al., 2012). Surprisingly, less work has been performed with attention cuing, specifically spatial attention, independently of multisensory integration. Given that multisensory integration and spatial attention orienting are different (although not exclusive) processes, I anticipated the involvement of different neural regions when using the looming sound as a cue.

Indeed, multisensory integration and crossmodal spatial attention cuing activate different neural networks. While the multisensory integration of the looming property activated the right claustrum/insula, amygdala, and bilateral occipital areas (Cappe et al., 2012), the auditory looming cue for shifting visuo-spatial attention activated frontal and parietal areas (see Chapter 5, Figure 18). Previous work has demonstrated a fronto-parietal network for shifting visuo-spatial attention (Corbetta and Shulman, 2002; Thiel et al., 2004; Corbetta et al., 2008). The parietal activation is presumably due to the fact that these areas are involved in establishing spatial awareness. Localizing objects in space (Bushara et al., 1999; Zatorre et al., 2002) and orienting spatial attention to locations in space activates parietal cortices (Sturm and Willmes, 2001; Farah et al., 1989; Gitelman et al., 1999). In addition, the frontal areas presumably take the role of controlling when to shift attention to locations of interest (Miller and Cohen, 2001).

Previous work has shown a preferential activation of the temporo-parietal junction (TPJ) for looming sounds (Seifritz et al., 2002; Neuhoff et al., 2002; Bach et al., 2008, 2015). However, the source estimation analysis for the looming cue benefit (see Chapter 5, Figure 18) did not show activity in the TPJ, but in the precuneus of the superior parietal lobe (SPL) instead. This is unsurprising given that I investigated the crossmodal cuing effect on visual processing and not the

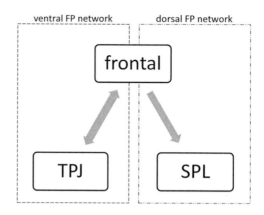

Figure 1: Schematic illustration of the fronto-parietal (FP) attention networks. This simplified schematic is adapted from Corbetta et al. (2008) and explains the interaction of the dorsal and ventral FP attention network. Frontal areas of the dorsal system control which stimuli can capture attention involuntarily, reflected by parietal activity in the temporo-parietal junction (TPJ). This information is communicated through frontal areas to the dorsal network which re-orients attention voluntarily, reflected by parietal activity in the superior parietal lobe (SPL).

processing of the sound itself. Nevertheless, both parietal areas feature in the FP spatial attention network. This is because there are two distinct FP attention networks: a ventral and a dorsal system as illustrated in Figure 1 (Corbetta and Shulman, 2002; Corbetta et al., 2008; Macaluso, 2010). The ventral system represents the involuntary capture of attention through behaviorally relevant stimuli. It typically activates a network of ventral frontal cortex and TPJ when salient stimuli, which are outside the current focus, capture attention. The dorsal system controls attention in a goal-driven way and activates a network of frontal eye field (FEF) and SPL. Both FP networks interact to reorient spatial attention. The preferential TPJ activation for looming sounds suggests stronger exogenous capture of attention for looming than comparable sounds. The activation of medial frontal gyrus (MFG) for the involuntary and voluntary attention might reflect the ventral system communicating the region of interest, where attention was captured involuntarily, to the dorsal system. The dorsal system then shifts attention voluntarily to this region of interest which manifests in SPL activation. Hence, the increased SPL activation for the looming cue benefit that I found may reflect the increased urge to shift attention to the cued location.

The activation of fronto-parietal areas gave rise to the auditory looming cue benefit for processing visual targets. This benefit in shifting spatial attention can be observed in the ERP response to the visual target. It manifests around 280 ms post visual target onset as a positive peak for the cue benefit (see Chapter 5, Figure 17). If activity related to visual processing is not removed by subtraction, the cue benefit appears as a positive plateau in the rise of the P3b. Interestingly, previous ERP work on crossmodal spatial attention shifting also illustrates this plateau in endogenously cued visual stimulus processing (Eimer, 2001, top of Fig. 2 and Fig. 4). This suggests that the peak identified in Chapter 5 (rP280), in fact, represents the benefit of shifting visuo-spatial attention. Given that this peak was larger for looming sounds, auditory looming cues might shift spatial attention more efficiently to peripheral visual targets, which is also reflected in faster response times to looming cued targets.

1.4.3 *Strategic allocation of attention through auditory cues (Chapter 4-6)*

Auditory cues are not only able to alert and orient attention to locations of interest but are also a helpful tool in scheduling tasks. That is, they can indicate when attentional resources should be dedicated to which task. Chapters 4, 5, and 6 of this dissertation address this functionality of auditory cues within a dual-task paradigm. Dual-tasking, especially when one task is continuous, requires timely switching between two tasks which is controlled by the executive system (Petersen and Posner, 2012; Dosenbach et al., 2008; Miller and Cohen, 2001). By setting out a goal at the beginning of a task, attention is devoted voluntarily to its best performance. In the case of Chapter 4-6's experiments, the participants' goal was to achieve their optimal visuo-motor tracking performance. Nevertheless, they were also required to identify occasional peripheral targets that could be preceded by an auditory cue at the target location. These transient signals (c.f., cues) captured attention involuntarily and indicated a necessary task switch. Hence, voluntary sustained attention for performing the continuous tracking task, as well as involuntary transient attention to direct attention to the periphery are required to interact to perform both tasks. The interaction of these two systems is controlled in a top-down fashion by the executive system that determines whether attention can be captured exogenously and schedules the switching between tasks.

Both, voluntary and involuntary attentional processes compete for the same limited attentional resources. It has been shown that attentional manipulations, such as the introduction of a central task, which requires attentional resources for processing, can eliminate exogenous spatial cuing effects (Santangelo and Spence, 2007). This could be due to attentional resources being devoted voluntarily to a task which reduces the amount of resources that auditory cues can capture exogenously. Increasing the demand of voluntary attention required to perform a task (e.g., by increasing task difficulty) has been shown to decrease the processing of auditory events (e.g., Wickens et al., 1983; Scheer, 2017). However, the relevance attributed to auditory events can also influence their processing (e.g., Dehais et al., 2014; Scheer, 2017). That is, Scheer (2017) found that sounds were processed more when they were relevant compared to when they were irrelevant to the task. A possible explanation for this might be that more attentional resources were made available for auditory processing in situations where auditory events were relevant (Talsma et al., 2006; Keitel et al., 2013).

CAN TASK-RELEVANT AUDITORY CUES SUPPORT TASK-MANAGEMENT? In this dissertation, auditory cues employed to capture attention exogenously were relevant to the task performance since they could be indicating a visual target. Hence, the executive system was required to balance the use of attentional resources for the voluntary attention task in the center and to ensure sufficient remaining attentional resources to enable involuntary attention capture in the periphery. Instead of investigating the processing of auditory cues in a dual-task paradigm (see, for example, Scheer, 2017), I was interested in how well these cues were able to draw attention away from the ongoing central task to occasionally perform the peripheral task. Interestingly, my auditory cues not only improved performance in the peripheral task, but also allowed participants to keep their eyes on the central task instead of looking at the visual targets. This suggests that auditory cues can help to shift visuo-spatial attention to the periphery, as well as make task switching more efficient. The cue prepares for the arrival of a target, frees attentional resources temporarily, directs attention covertly to the prospective target location, and as soon as this 'interrupting' task is performed, switches back to the ongoing task. In contrast, an uncued peripheral target is surprising and since no covert attention can be directed to the target location in advance, overt attention is directed to the peripheral task to compensate for not attending. This can decrease central task performance given that the eyes are not directed towards this task.

Overall this has significant practical implications in that auditory cues are not only helping in detecting unnoticed visual targets but also in managing the strategic allocation of attention such that both, central and peripheral, tasks can be performed as best as possible. In the context of driving, this would suggest that driving performance is better when external events are cued compared to having the drivers detect them on their own, possibly at a later point in time, because they might have to take their eyes off the road.

1.4.4 *Conclusions and outlook*

In this dissertation, I have been able to show that auditory cues are useful for managing attention. More specifically, I have demonstrated that auditory cues such as verbal commands and auditory icons can alert by mobilizing attentional resources. However, the findings suggest that these two types of cues should be employed complementary depending on the intended goal. Based on the brain responses to these auditory cues, verbal commands seem to be more appropriate for urgent notifications while auditory icons should be used for updating of changing environment conditions.

Furthermore, I showed that auditory looming cues improve orienting of visuo-spatial attention to relevant locations while performing another cognitive demanding task. This benefit was especially apparent at the looming sound's off-set that signals the arrival of an approaching object. The improved performance to looming cued visual targets was mirrored by the neural response of the cue benefit, ERP component rP280. A network of frontal and parietal brain regions, involved in spatial reorienting, seems to underlie this preferential processing of looming cued targets.

The third aspect involved in managing attentional resources is the scheduling of task switches. Here, I found auditory cues to be helpful in indicating an upcoming task switch. Auditory cues allowed participants to shift attention in advanced to the periphery without moving the eyes. Moving the eyes away from the central task might come at the cost of decreased central task performance. Taken together, auditory cues are an efficient tool to manage the limited attentional resources available by alerting, directing attention to unnoticed information, and by managing the strategic allocation of attention between different tasks.

Although the work of this dissertation expanded the state of the art on the use of auditory cues for attention management, many questions remain to be answered.

Future work employing EEG can, for instance, investigate the preference of semantically comparable verbal commands such as *fuel* and *tank* for notifying the driver to get gas. In addition, future work can also examine how auditory icons are processed if they do not match the environmental circumstances by evaluating the brain responses to these auditory notifications.

The beneficial use of auditory looming cues for orienting visuo-spatial attention opens a new area of research. Future work making use of looming sounds as orienting cues can, for example, investigate how long transient attention can be sustained by varying the duration of the auditory looming cue. Furthermore, future studies can investigate the connectivity of frontal and parietal areas, identified here, to address how these regions might influence each other to give rise to the looming cue benefit.

This dissertation consists of a collection of four manuscripts that are either published, submitted for publication, or in preparation to be published. The following presents more details on the contribution of the different authors to these manuscripts.

The candidate developed the conception for all the studies, including the experimental design as well as the software to carry out the experiments. The data analysis and interpretations were entirely by the candidate's own work.

The candidate is the first author on all four manuscripts while she shares equal authorship with Lewis L. Chuang for the first manuscript. The co-authors Lewis L. Chuang and Heinrich H. Bülthoff supervised the work and assisted in the manuscripts' revision. Stas S. Krupenia supported the engineering aspects of vehicle simulation that features in the first two manuscripts. Makoto Miyakoshi advised in the data analysis of the fourth manuscript.

1. Chuang L. L., Glatz C., and Krupenia S. S. (2017). Using EEG to understand why behavior to auditory in-vehicle notifications differs across test environments, 9th International Conference on Automotive User Interfaces and Interactive Vehicular Applications (AutomotiveUI '17), ACM Press, New York, NY, USA, 123-133.

2. Glatz, C., Krupenia, S. S., Bülthoff, and H. H., Chuang, L. L. (2018). Use the right sound for the right job: Verbal commands and auditory icons for a task-management system favor different information processes in the brain. In Proceedings of the 2018 CHI Conference on Human Factors in Computing Systems. ACM, New York, NY, USA.

3. Glatz, C. and Chuang, L. L. (2018). The time course of auditory looming cues in redirecting visuo-spatial attention (submitted).

4. Glatz, C., Miyakoshi M., Bülthoff, H. H., and Chuang, L. L. (2018). rP280: An ERP marker for the looming cue benefit and its underlying neural networks (in preparation).

Parts of this work were also presented at the following conferences:

- Glatz C., Bülthoff H. H., and Chuang L. L. (2015). Sounds with time-to-contact properties are processed preferentially, 57th Conference of Experimental Psychologists (TeaP 2015), Hildesheim, Germany.

- Glatz, C., Bülthoff, H. H., and Chuang, L. L. (2015). Warning Signals With Rising Profiles Increase Arousal, Human Factors and Ergonomics Society Annual Meeting (HFES 2015), Sage, London, UK, 1011.

- Glatz, C., Bülthoff, H. H., and Chuang, L. L. (2015). Attention Enhancement During Steering Through Auditory Warning Signals, Workshop on Adaptive Ambient In-Vehicle Displays and Interactions in Conjunction with AutomotiveUI 2015 (WAADI'15), 1-5.

- Glatz C., Bülthoff H. H., and Chuang L. L. (2016). Looming warnings orient and sustain attention at cued location, 50. Kongress der Deutschen Gesellschaft für Psychologie (DGPs 2016), Leipzig, Germany.

- Glatz C., Bülthoff H. H., and Chuang L. L. (2016). Why do Auditory Warnings during Steering Allow for Faster Visual Target Recognition?, 1st Neuroergonomics Conference: The Brain at Work and in Everyday Life, Paris, France.

- Borojeni S. S., Chuang L. L., Löcken A., Glatz C., and Boll S. (2016). Tutorial on Design and Evaluation Methods for Attention Directing Cues, In: Adjunct Proceedings, 8th International Conference on Automotive User Interfaces and Interactive Vehicular Applications (AutomotiveUI '16), ACM Press, New York, NY, USA, 213-215.

- Glatz C. (2017). Auditory warnings for steering environments, 3rd Berlin Summer School Human Factors, Technische Universität Berlin: Zentrum für Mensch-Maschine-Systeme, Berlin, Germany, 19-20, Series: MMI-Interaktiv, 17.

- Glatz C., Ditz J., Kosch T., Schmidt A., Lahmer M., and Chuang L. L. (2017). Reading the mobile brain: from laboratory to real-world electroencephalography, 16th International Conference on Mobile and Ubiquitous Multimedia (MUM 2017), ACM Press, New York, NY, USA, 573-579.

- Lahmer M., Glatz C., Seibold V. C., and Chuang L. L.(2018). The looming benefit in driving with ACC, Submitted to the 2nd Neuroergonomics Conference: The Brain at Work and in Everyday Life.

- Löcken A., Borojeni S.S., Müller H., Gable T. M., Triberti S., Diels C., Glatz C., Alvarez I., Chuang L. L., and Boll S. (2017). Towards Adaptive Ambient In-Vehicle Displays and Interactions: Insights and Design Guidelines from the 2015 AutomotiveUI Dedicated Workshop, 325-348. In: Automotive User Interfaces: Creating Interactive Experiences in the Car, (Ed) G. Meixner, Springer, Cham, Switzerland.

2

INDIVIDUAL DIFFERENCES IN RESPONDING TO AUDITORY CUES

This chapter has been reproduced from an article published at Automotive'UI 2017: Chuang, L. L., Glatz, C., and Krupenia, S. S. (2017). Using EEG to understand why behavior to auditory in-vehicle notifications differs across test environments. In Proceedings of the 9th International Conference on Automotive User Interfaces and Interactive Vehicular Applications, pp. 123-133. ACM, New York, NY, USA. DOI: http://dx.doi.org/10.1145/3122986.3123017

2.1 ABSTRACT

In this study, we employ EEG methods to clarify why auditory notifications, which were designed for task management in highly automated trucks, resulted in different performance behavior, when deployed in two different test settings: (a) student volunteers in a lab environment, (b) professional truck drivers in a realistic vehicle simulator. Behavioral data showed that professional drivers were slower and less sensitive in identifying notifications compared to their counterparts. Such differences can be difficult to interpret and frustrates the deployment of implementations from the laboratory to more realistic settings. Our EEG recordings of brain activity reveal that these differences were not due to differences in the detection and recognition of the notifications. Instead, it was due to differences in EEG activity associated with response generation. Thus, we show how measuring brain activity can deliver insights into how notifications are processed, at a finer granularity than can be afforded by behavior alone.

2.2 INTRODUCTION

Notifications are a fixture of in-vehicle environments. They are designed to direct users, who would be engaged otherwise, to aspects of the environment that require a response (e.g., fuel indicator lights). Recent advances in automated driving will increase the importance of notifications, especially when the duties of the human operator transition from vehicle control to vehicle supervision (Bainbridge, 1983; Casner et al., 2016). Indeed, research on the design of in-vehicle interactions have rapidly shifted, in recent years, towards addressing the anticipated user requirements of automated vehicles (Kun et al., 2016). Given the rapid pace of innovation in technology and design, it is surprising that we continue to have few tools at our disposal that allow us to truly appreciate how users process and act upon in-vehicle notifications. Here, we combine the analysis of behavioral responses with electroencephalography (EEG) recordings to better understand how auditory notifications were processed for information and responded to across different test environments and participant groups.

The design space of notifications is large. This gives rise to infinite variations of how in-vehicle notifications ought to be designed and for which purpose. While guidelines have been proposed for the design of in-vehicle displays (e.g., Green et al., 1995; Ross et al., 1996; Stevens et al., 2002), they tend to be based on studies with a focus on critical-safety behavior. Ultimately, notifications are implemented according to whether or not they will be effective in safely eliciting the desired responses, in the environment for which they were designed for. Unfortunately, evaluating notifications by performance measures alone can only reveal whether a given implementation is better or worse than its control comparison. In order to understand why a given notification results in better or worse performance than originally anticipated, it is necessary to evaluate the extent to which the notification is perceived, processed, and elicits a response. For this, it is necessary to inspect how the brain responds to notifications.

Methods for neuroimaging are becoming more accessible. Recent developments in neuroimaging technology, especially with regards to EEG, have focused on the ease of application and user mobility (Gramann et al., 2014). In spite of this, valid concerns persist with regards to their suitability for use in realistic test environments, especially since the presence of electronic devices can introduce substantial noise into EEG recordings. On a more practical note, EEG measurements are often expected to impose implementational costs on the researcher that might be deemed unnecessary, especially when behavioral and self-report measures are expected to suffice.

Nonetheless, the time-varying EEG signal offers a detailed inspection of how information is processed by the human user, which cannot be achieved with performance measurements alone. With regards to the evaluation of notifications, EEG measurements can reveal how the brain automatically detects and consciously identifies notifications. They can also indicate how the brain prepares itself to generate an appropriate response, pending the identification of the notification. Performance measurements implicitly treat the human operator as a single stimulus-response unit and do not, in themselves, distinguish between perception, cognition, and action.

Our research aim was to demonstrate that employing EEG methods can allow us to account for why behavioral responses to auditory notifications might differ across different instances of testing. Specifically, between professional drivers tested in a high fidelity driving simulator and students tested in a psychophysical laboratory. This is an experience that is common to many researchers when evaluating novel designs of notification interfaces for deployment in the "real world". Often, interfaces are first prototyped and evaluated under highly controlled conditions before they are deployed in more realistic environments and tested with their intended users. When the behavioral results of a highly controlled test do not generalize to a more realistic one, it is often difficult to establish the reasons that might have caused this.

Here, we show that EEG measurements can complement behavioral results to provide a better resolution for understanding how notifications are processed by users across different settings. Unlike behavioral performances, the appropriate application of EEG measurements allows the researcher to discriminate how notifications are processed by the brain across at least three different stages of information processing, namely perceptual, cognitive, and response stages (Wickens, 2002, 2008). In this regard, it offers researchers the ability to investigate the outcome of information processing at the various stages that lie in between the presentation of a stimulus (i.e., auditory notification) and a behavioral response (i.e., keypress).

We report two experiments that presented participants with identical tasks but under two different test environments. All of our participants were required to respond to auditory notifications, which were previously designed to direct the attention of commercial truck drivers to the occurrence of task requirements during a long distance, automated vehicle mission (Fagerlönn et al., 2015). They were also presented irrelevant distractor sounds, which they had to ignore, and a dynamic visual scene, which varied in its realism according to the test environment. The first experiment was performed as a highly-controlled psychophysical experiment (N=15), with low mission fidelity, and on young and untrained student participants. The second experiment was performed on professional truck drivers in a high-fidelity driving simulator (N=14). Our findings are as follows:

1. professional drivers in a high fidelity simulator were slower and less sensitive in discriminating target notifications from distractor sounds

2. the EEG activity for notification detection and identification did *not* discriminate between the two test environments

3. the EEG activity for correct responses to the notifications (i.e., Bereitschaftspotential; BP) discriminated between the two test environments

4. thus, we attributed observed behavioral differences to differences in the sample demographic and not to differences in the test environment.

2.3 RELATED WORK

In-vehicle notifications are often designed to shift user attention from the primary task of engagement (*e.g.* driving) to a critical event (*e.g.* low fuel). With advances in vehicle sensing, notifications can also be designed to direct a user's attention to safety-critical aspects of the driving task (*e.g.* pedestrian detection; Tsimhoni and Flannagan, 2006; Zhang et al., 2016) or to support decision making (*e.g.* lane-changing; Löcken et al., 2015). With the increased adoption of automated vehicles, we expect the role of notifications to grow in prominence. Besides takeover notifications that prompt users to resume vehicle handling (*e.g.* Borojeni et al., 2016; Politis et al., 2015), we also expect task-management notifications to pervade the in-vehicle environment as the scope of permissible in-vehicle activities grows (Paul et al., 2015). In particular, commercial vehicles (i.e., trucks) will stand to benefit from the effective introduction of task-management notifications. This is because commercial drivers might be expected to assume additional responsibilities such as delivery logistics, as their responsibility for vehicle diminishes with increasing handling automation.

2.3.1 *Auditory notifications*

Auditory displays tend to be a favored delivery medium for notifications, given that they are not in direct conflict with the visuo-motor demands of vehicle handling (Nees and Walker, 2011). Auditory information that is presented during driving has been claimed to be more deeply processed than visual information, given that it is more likely to be recalled post testing (Mollenhauer et al., 1994). However, it can also be perceived as being more distracting. In any case, auditory notifications inhabit a large design space with manipulable parameters, which include their formant, duration, interpulse interval, onset/offset latency and more. This allows them to be flexibly tuned in order to communicate information, such as urgency (Edworthy et al., 1991), even whilst being moderated for undesirable side-effects, such as perceived annoyance (Marshall et al., 2007).

Some sounds are more effective than others, particular when operational concerns are taken into consideration. For example, a driving simulator study compared different auditory warnings for signaling potential headway collision and found that the sound of a car horn and a tone with a looming time-to-contact intensity resulted in the best braking latencies (Gray, 2011). Nonetheless, the sound of a car horn also resulted in more unnecessary braking responses than the looming tone. A separate study compared four classes of auditory notifications—namely, abstract sounds, auditory icons, nonspecific environment sounds, and speech messages—for their efficacy in cuing for driving situations such as low tire pressure, low oil level, engaged handbrake whilst driving, and others (McKeown and Isherwood, 2007). This study found that speech messages and auditory icons generated faster and more accurate responses than either environment sounds or abstract sounds.

In these mentioned examples, auditory displays were often evaluated on the basis of behavioral responses—namely, accuracy and response latencies. Although brain responses offer a finer granularity of information processing, they are rarely employed in the evaluation of auditory notifications.

2.3.2 Event-related potentials (ERPs)

EEG refers to the measured electrical activity of surface electrodes placed on the human scalp (typically in the range of $10–100\mu V$), which can be attributed (in part) to brain activity (Schomer and Da Silva, 2012). Event-related potentials (ERPs) are changes within a pre-specified time window of EEG activity that are generated after or prior to a known event. In this work, we focus on two physical events: (i) the presentation of a target notification, (ii) the participants' self-generated response to target notifications. Respectively, this allows to first understand how our participants' brains *detect* and *recognize* the presented notification and, next, *decide* to generate a behavioral response relative to their self-generated responses. This corresponds to three stages of information processing that lie between presenting a notification and generating a response.

Stimulus ERPs for auditory events

Auditory stimuli are frequently employed in ERP studies as they elicit waveforms with identifiable components that are associated with cognitive mecha-

Figure 2: Left: A student participant in a psychophysical laboratory (Department for Human Perception, Cognition, and Action, MPI for Biological Cybernetics, Tübingen, Germany). Right: A professional commercial driver in a truck driving simulator (Styling and Vehicle Ergonomics, Scania CV AB, Södertälje, Sweden)

nisms. Of current interest are the slow transient responses that arise from the auditory and associated cortices (i.e., 50 msec after sound onset). A popular paradigm (i.e., oddball paradigm) presents two discriminable sounds, one more frequently than the other (e.g., 80% to 20%). Participants are only required to respond to the infrequent sound, which are termed targets. This corresponds to a real-world scenario where auditory notifications have to be detected and identified against an auditory background of distractors. Subtracting the EEG activity generated by the frequent distractor sounds from that generated by targets results in a difference waveform with two interpretable components. First, the mismatch negativity (MMN), which is an early negative deflection with a typical latency of about 140 msec. The MMN is associated with an automatic process that responds in proportion to the perceived deviance of the targets from the distractors. It is generated even when there is no task involved. Second, the P3 ,which is a positive late deflection with a latency between 450–600 msec. The P3 is only generated if the subject is attending to the stimuli in a way that demands a response (but, see Scheer et al., 2016). Working memory processes that underlie context updating are claimed to be represented by the P3 component (Donchin and Coles, 1988). Indeed, fMRI studies have localized brain regions, which are typically implicated with conscious effort and working memory processes (i.e., insular cortex, inferior parietal and frontal lobes), as generators of the P3 (*e.g.* Linden et al., 1999). In the current study, we evaluated MMN and P3 responses to target auditory notifications in order to determine whether behavioral differences across different test environments reflected changes in their automatic detection and/or voluntary identification.

Response ERPs for motoric responses

Brain responses can also be measured prior to response actuation. When EEG activity are temporally aligned to the generated responses of measured individuals, it is possible to observe a slow change in potentiation, leading up to the response onset. This observation was first reported in 1964 and was termed the *Bereitschatpotential* (i.e., readiness potential; BP; Kornhuber and Deecke, 1964; Shibasaki and Hallett, 2006).

The BP is maximal at the midline centro-parietal area (i.e., CPz). The adoption of a common average reference, such as in the current work, means that it is observed as a positive potential shift in the parietal and occipital electrodes and as a negative potential shift in the frontal electrodes. In an interesting series of experiments, recorded participants were asked to report the time when they decided to generate volitional keypress responses (Libet, 1985). Here, the BP was found to be initiated approximately 350 msec prior the participants' reported times, which raised philosophical doubts on the nature of *free will*. Putting such existentialist concerns aside, most researchers agree that the initiation of BP has its physiological origins in the supplementary motor area (SMA), a brain region that is implicated in the generation of motor responses, at least in the case of hand movements (Praamstra et al., 1996). Thus, BP could be regarded as a cortical decision to initiate a motor action (prior to the conscious realization of the decision itself!). More recently, it has been claimed that this cortical decision can be volitionally cancelled up till a point (i.e., 200 msec prior to response), after which the generation of a motor response inevitable (Schultze-Kraft et al., 2016). For our current purposes, we treat BP as an indicator for the cortical decision to respond, which is initiated earlier than the recorded response itself. This allows us to determine the latency between the cortical decision and the actuated response, as well as the amplitude of the cortical decision itself which is not evident in a binary key- or button-press response.

2.4 STUDY

This study was a between groups design that compared behavioral performance and brain responses to auditory notifications across two different experimental settings: (i) a highly controlled psychophysical laboratory, (ii) a high fidelity driving simulator environment. The whole experiment took approximately 2.5 hours to complete, including preparation time and debriefing.

2.4.1 Participants

Thirty participants—that is, fifteen undergraduate students (mean age(*std*)=26.1(*4.0*) years; 9 males) from the University of Tübingen, Germany, and fifteen professional commercial drivers (mean age(*std*)=41.4(*12.1*) years; 13 males) who were employees of Scania CV AB, Sweden —performed the task reported here. The data of one professional driver, from the driving simulator testing, had to be excluded from further analysis because only one response was recorded throughout the entire experiment. Besides demographic differences, the primary difference between these two groups was the test environment that they experienced (see 2.4.2). All participants reported no known hearing defects and provided signed informed consent.

2.4.2 Apparatus and Stimuli

Psychophysics laboratory

The psychophysics laboratory had black walls, was insulated for external sounds, and had an ambient sound level of approximately 40 dB (Figure 2, left). Visual stimuli were presented on a desktop display (60° field-of-view; 45 cm distance to fixed chin-rest). The visualization was rendered by customized software written in Matlab R2013b (The Mathworks, Natick, MA). The visualization consisted of a cursor that drifted horizontally between two vertical lines, which represented a vehicle's position in a single lane. Participant responses were collected via keypresses on a standard USB keyboard. Sound presentation was controlled by an ASIO 2.0 compatible sound card (SoundBlaster ZxR; Creative Labs) and displayed via stereo speakers, each placed on either side of the display.

High fidelity driving simulator

The driving simulator was designed to simulate the operation of commercial truck vehicles. Participants sat in a realistic cabin interior, based on an existing truck seating buck that consisted of a pneumatic seat, a steering column complete with wheel and shaft, instrument cluster, and the remaining dashboard (Figure 2, right). Visualization was presented on a front-projection three wall display (approx. 150° field-of-view; 450 cm distance to head), and via two pairs of 2 vertically-aligned displays each, attached to either side of the cabin, that simulated side rear-view mirrors. The visualization was rendered by a customized graphical engine (i.e., VISIR) that created 3D environments from OpenDrive8 road network files (xodr) and from additional landscape description file (xml). Here, we presented a highway scene from Linköping and Norrköping, with two lanes for congruent traffic and two lanes for opposing traffic. The participant inhabited the far-right lane. The highway was populated with low traffic density, including the occasional headway vehicle. Experimental responses were collected via dedicated buttons that were located on the steering wheel. Sound presentation was controlled by an ASIO 2.0 compatible sound card (RME HDSP 9632; RME Intelligent Audio Solutions) and displayed via a 5.1 surround sound system, installed around the driver's seat.

Experimental Stimuli

Twelve target and ninety distractor sounds were used in this experiment. These were modified from sounds that were designed as part of a project (MODAS: Methods for Designing Future Autonomous Systems) to cue truck drivers to attend to possible non-driving tasks (Fagerlönn et al., 2015; Krupenia et al., 2014). All sounds had a duration of 500 msec.

There were two notifications for each of six non-driving tasks. They were a verbal command in Swedish and an auditory icon (in brackets): (i) system (synthetic tone), (ii) convoy (train whistle), (iii) driver (human whistle), (iv) weather (raindrop), (v) road (rumbling), (vi) traffic (car horn). The distractor sounds were random permutations of four sounds, two verbal commands and two auditory icons, played simultaneously in reverse.

EEG recording

EEG was recorded on a dedicated PC using a 59-channel active electrode array that was affixed to the scalp using an elastic whole head cap, which specified for pre-determined sites including those corresponding to the international 10-20 system (ActiCap System, Brain Products GmbH, Munich, Germany). The horizontal and vertical electrooculogram (HEOG/VEOG) were recorded with four electrodes attached to the right and left canthi as well as above and below the left eye. FCz was used as an online reference for all channels. Prior to testing, electrode gel was applied to ensure that electrode impedance was $< 20k\Omega$ for each channel. EEG signals were digitized at a rate of 1000 Hz. EEG recordings

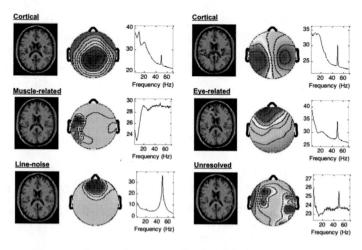

Figure 3: Six examples of clusters of dipoles (blue) and their mean position (red), their projected scalp activity, and power spectral density (inset: left to right), derived from EEG recordings in the driving simulator. First row: Cortical dipoles that are likely to be associated with auditory processing (left) and motor response generation (right) respectively. Second row: Non-cortical dipoles that are associated with muscle activity (left) and eye-movements and -blinks (right). Third row: Non-cortical dipoles that are due to electrical line noise (left) and unresolved variance in EEG recording (right).

were synchronized with experimental events via a parallel port connection to the experimental PC.

2.4.3 Task

All participants performed the same task, regardless of the experiment setting. They were required to attend to the visual scene throughout the experiment. Participants were informed that this was a stimulated automated driving scenario and that no steering was necessary. Whenever they heard a sound, they were required to respond if it was a target notification and to ignore it if it was a distractor sound. The inter-stimulus interval was randomly selected from a uniform distribution across 1800-2000 msec. Participants could respond within 2000 msec of the onset of the target notification. Failures to do so were considered misses. Responses to distractor sounds were treated as false alarms. Each experiment presented approximately 980 sounds in total. Of these, 20% were target notifications.

2.4.4 Results: Behavioral performance

Our participants' performance were assessed in terms of the median of their correct response times (RT) and discriminability index (i.e., d'). The discriminability index is calculated as the difference between the z-scores of hit and false-alarm rates, whereby hits and false-alarms were responses to target and distractor sounds respectively (Macmillan and Creelman, 1991). Welch's t-tests for independent samples were performed to compare behavioral performance in the high fidelity driving simulator and the psychophysical laboratory (Delacre et al., 2017). The adopted criteria for statistical significance was $\alpha = 0.05$.

Participants in the high fidelity driving simulator were slower (mean=1262 msec; std.=120 msec) in their correct responses than those in the psychophysics laboratory (mean=1062 msec; std.=139 msec). This difference (mean=200 msec) is statistically significant ($t(26.8)$=4.18, $p < 0.001$, Cohen's d=1.54) and has a 95% confidence interval from 102 to 299 msec.

Participants in the high fidelity driving simulator were less sensitive (mean=4.37; std.=1.03) in discriminating the auditory target notifications from their distractor counterparts, than those in the psychophysics laboratory (mean=5.25; std.=0.60). This difference (mean=0.88) is statistically significant ($t(20.6)$=2.81, $p < 0.05$, Cohen's d=1.06) and has a 95% confidence interval from 0.23 to 1.54.

To summarize, behavioral results indicated slower and less discrimination sensitivity for auditory notifications in the driving simulator, compared to the psychophysics laboratory.

2.4.5 Results: EEG/ERP responses

Data collection, signal processing, and statistical analysis

Data pre-processing and analysis was performed offline with Matlab (The Mathworks, Natick, MA) scripts based on EEGLAB v.14[1], an open source environment for processing electrophysiological data (Delorme and Makeig, 2004). The following steps were performed on EEG data prior to analyzing the ERPs of stimuli and responses (Bigdely-Shamlo et al., 2015). First, the data was downsampled

---

[1] https://sccn.ucsd.edu/eeglab/

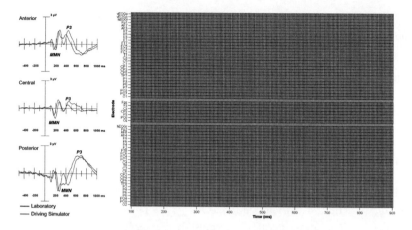

Figure 4: LEFT: Stimulus ERPs are illustrated by and labelled in three difference waveforms that depict averaged EEG activity of electrode groups from anterior, central, and posterior regions. RIGHT: MUA results plot statistically significant t-values between the two participant groups for every electrode and time-point. The analysis reveals that there are no statistically significant electrode-time regions that proceed the auditory notification.

to 250 Hz to reduce computational costs. Next, a high-pass filter (cut-off=0.5 Hz) was applied to remove slow drifts, 50 Hz electrical line noise from the environment was removed using the CleanLine algorithm, and bad channels were removed using the ASR algorithm. Next, all electrodes were re-referenced to their common average, and each participants dataset was separately submitted to an Adaptive Mixture ICA to decompose the continuous data into source-resolved activity (Delorme et al., 2012). On these learned independent components (IC), equivalent current dipole model estimation was performed by using an MNI Boundary Element Method head model to fit an equivalent dipole to the scalp projection pattern of each independent component. ICs whose dipoles were located outside the brain were excluded as well as those that had a residual variance of over 15%. Within each participant group, ICs were clustered into 30 clusters using k-means based on their mean power spectra, topography, and equivalent dipole location.

Figure 3 provides examples of dipole clusters with either cortical (first row) or non-cortical origins (e.g., muscle and eye activity (second row), electrical activity from environment sources (third row)). Non-cortical components were identified on the basis of their power spectral density, scalp topology, and loca-

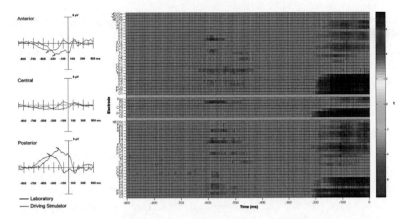

Figure 5: LEFT: Stimulus ERPs are illustrated by three waveforms that depict averaged EEG activity of electrode groups from anterior, central, and posterior regions. The BP peaks are indicated by arrows. RIGHT: MUA results reveal statistically significant differences between driving simulator and laboratory recordings in two time periods (i.e., 600-430 and 220-0 msec before responses).

tion in a volumetric brain model (Jung et al., 2000). As might be expected, there were more non-cortical dipole components found in participants who performed the experiment in the driving simulator (N=15) than those from the psychophysical experiment (N=14). In other words, EEG recordings were contaminated by the activity of more non-cortical components in the driving simulator environment than in the psychophysical laboratory. Non-cortical dipole clusters were removed from the EEG recording and the remaining EEG activity was subjected to comparative analysis for the two participant groups.

Specifically, we derived a stimulus and a response ERP for each participant. This was achieved by mean-averaging the EEG activity of a time-window (also termed an epoch; 1000 msec after/before the relevant trigger event, baselined to 500 msec before/after the stimulus/response), across all epochs. Stimuli ERPs were defined by the differences in EEG responses to target notifications and distractor sounds, prominent components are MMN and P3, which are respectively associated with the information processing aspects of notification detection and identification (Picton, 2011). Response ERPs were defined by EEG activity prior to the registration of our participants' responses. It is defined by a single component, which manifests itself as a single peak that changes from negative to positive polarity, from the anterior to posterior electrodes. The left panels of Figures 4 and 5 illustrate the averaged activity of anterior (Fz, F1, F3, F5, F2, F4,

F6, FC1, FC3, FC5, FC2, FC4, FC6), central (Cz, C1, C3, C5, C2, C4, C6), and posterior (Pz, POz, Oz, P1, P3, P5, P2, P4, P6, PO3, PO7, PO4, PO8, O1, O2) electrodes for stimulus and response ERPs respectively. The profile of the derived waveforms were consistent with our expectations for both test environments.

To evaluate our EEG recordings for differences across the two test settings, we performed separate mass-univariate analyses (MUA) for the stimuli and response ERPs (Groppe et al., 2011). This allowed us to determine the time points of individual electrodes that were statistically significant for waveform differences between the EEG activity recorded across the two test environments. False discovery rate control was applied (i.e., FDR-BH; Benjamini and Hochberg, 1995). We illustrate MUA results as raster plots of electrode channels across time (see Figures 4 and 5, right), whereby statistically significant t-values are represented by color intensity. The raster plots consist of three panels whereby the top and bottom panels indicate right- and left-hemispheric electrodes respectively, and the middle panel indicate mid-line electrodes. Within each panel, electrodes are vertically ordered from anterior to central to posterior electrodes.

Stimulus ERPs

The MUA results (Figure 4, right) reveal no statistically significant differences between the difference waveforms (Figure 4, left) of participants from the laboratory (–) and driving simulator (–) test settings. This suggests that the notifications elicited equivalent brain responses for detection and identification in both groups of participants, regardless of their test environments.

Response ERPs

The MUA results (Figure 5, right) reveal statistically significant differences, particular in the frontal (e.g., Fpz, Fz) and posterior (e.g., Pz, POz, Oz) electrodes. To understand these differences, the reader should recall from the behavioral results that truck drivers in the driving simulator generated slower responses than students in the psychophysical laboratory, of about 200 msec. We note that the time periods of significantly different EEG activity are of similar duration (i.e., 600-430 and 220-0 msec before the response event). This means that the time that it took for truck drivers in the driving simulator to generate a behavioral response after the initiation of corresponding brain activity (i.e., BP) was approximately 220 to 230 msec longer than it took for the undergraduate students in the psychophysical laboratory. In addition, the peak amplitudes of the BP in the frontal and posterior electrodes were smaller in the truck drivers than the undergraduate students (Figure 5, left).

2.5 DISCUSSION

What inferences can we draw when auditory notifications, which were designed in the confines of a well-controlled laboratory, elicit different behavioral responses in more realistic settings? The current study demonstrates that EEG measurements can provide some clarity when the explanatory power of behavioral responses are limited.

In this work, we found statistically significant behavioral differences across two test settings. Professional drivers were slower and less sensitive in detecting target notifications in a high fidelity driving simulator compared to student participants tested in a psychophysical laboratory. To begin, this is surprising for at least two reasons. First, the verbal commands were in the native language of the professional drivers. Second, the professional drivers understood what these notifications represented in the context of their jobs. Thus, we might have assumed that professional drivers to have responded faster and more accurately. Although the professional drivers were slower than our student volunteers by approximately 200 msecs. They continued to respond in an acceptable time range (i.e., less than 2 secs, the recommended time headway for preventing rear-end collisions). From this, the current auditory notifications might be considered to be suitable for fulfilling their intended function of task management.

Based on behavioral data alone, the worse performance of the professional drivers could be attributed to several reasons. For example, the driving simulator could have provided a more immersive environment that reduced the perceptual saliency of the notifications. Alternatively, it could have been due to age or motivational differences between the professional drivers and student participants. Last, but not least, performance differences could have resulted from technical differences in the auditory displays or input devices. This host of plausible interpretations often plague comparative user studies that rely solely on behavioral measurements.

With EEG measurements, we were able to infer that the behavioral differences that we observed were due to differences between the professional drivers and the student participants. Specifically, in how their brains prepare themselves prior to responding. Our reasoning is as follows. First, notifications were unlikely to have been detected or identified differently, given that the brain responses associated with these processing stages (i.e., MMN and P3 respectively) were similar across the two participant groups. In other words, the auditory notifications were robustly perceived across the two different test settings. Next, differences were found in the response ERPs. More specifically, the response ERPs for professional drivers had a longer latency than student participants between the BP peak and the recorded response. In other words, more time elapsed for the professional drivers between cortical decision-making to respond and the response itself. This difference in the latencies from BP initation to the motor response was about 220 msec, and gave rise to the statistical differences that the MUA analysis revealed. This difference in EEG activity between the two groups converges with the difference that we found with behavioral response times (i.e., 200 msec). Finally, the ERPs for response generation in professional drivers had smaller peak amplitudes than in student participants. This suggests that reduced cortical activity observed in the professional drivers prior to responding could have resulted in later responses. This possibility rules out alternative reasons for slow responding, such as the sub-optimal physical ergonomics of the truck cabin or the physical layout of the input device.

Taken together, the combination of EEG and behavioral measurements show that the current auditory notifications were sufficiently salient and robust across different test environments. Although professional drivers exhibited slower and less sensitive discrimination performance, it was not likely to be due to the notification design or the physical environment. Instead, it was due to differences in the sample demographics. Thus, subsequent effort in this scenario ought to be invested in understanding and mitigating for human factor limitations, rather than in further refinements in the design of notifications or the physical interface.

The current work is restricted to the presentation of auditory notifications. There are other channels of notification delivery that remain to be considered, such tactile notifications (e.g., vibrations) which have been claimed to be more easily discriminable and less interfering with the task of driving compared to auditory notifications (Cao et al., 2010).

To reiterate our main point, comparisons between different types of notifications and across different deployment settings can result in conflicting evidence, across independent studies and even within the same study. For example, while (Cao et al., 2010) reported that auditory notifications elicited shorter response times, a meta-analysis showed that tactile notifications elicited faster responses, at least for low-urgency messages (Lu et al., 2013). The same meta-analysis emphasized that moderating factors play a critical role in determining the suitability notification delivery and design. For example, tactile notifications might be more discriminable, but only if they have low-complexity (cf., Cao et al., 2010); responses to auditory notifications are more accurate for high-complexity information such as in the current study. This variability of empirical evidence across the diverse design and deployment space means that it is insufficient to simply focus on behavioral responses to user interfaces.

2.6 CONCLUSION AND OUTLOOK

The current paper contributes by demonstrating how EEG methods could allow us, as researchers, to identify the stage of information processing that results in differences in behavior. Such an approach will allow us to target the limitations of our designs for user interfaces more selectively and emphasize the aspects that are more deserving of our attention.

While EEG measurement is not a panacea, the current work shows that it can provide insight into how information is processed, at a finer granularity than behavioral responses alone. Furthermore, it can help to deliver insight when the deployment settings of our designs change across test phases. Here, it assured us that the designed notifications were sufficiently robust to be processed by the brains of their users, regardless of differences in the test environments.

This level of understanding, namely of how the information communicated by notification interfaces is processed by the brain, will be increasingly important especially as we attempt to increase the design space and functional diversity of notifications. Nowadays, notifications are designed, not only to capture the user's attention at all costs but, to be sensitive to the user's goals and requirements (McCrickard and Chewar, 2003). *Ambient notifications* represent a particular class of notifications that will be difficult to evaluate if behavioral measurements are all that we have to rely on. This is because *ambient notifications* are, by definition, designed to inform the user without eliciting behavior that would interrupt existing activity (Löcken et al., 2017; Pousman and Stasko, 2006). With such notifications, responses from the brain could be measured instead of behavioral responses.

The current work relied on high density EEG recording equipment. While the use of medical grade equipment can be feasibly implemented in a real car, and even for the actuation of emergency braking (Haufe et al., 2014), doing so might not be expedient for many researchers. A recent evaluation suggest that simpler and more convenient EEG setups, for example those that use dry electrodes, could be implemented in a vehicle environment at a reasonable signal to noise ratio (Zander et al., 2017). Innovations in electrode designs, such as an around-the-ear EEG placement (Bleichner et al., 2016), could further allow for brain responses to be measured without imposing on users the inconvenience of donning an unsightly EEG cap.

As we continue to innovate in-vehicle interfaces to keep up with the demands of user expectations, it is only appropriate that we also innovate our means for evaluating these interfaces to keep up with the demands of inferential rigor. The current work demonstrates the viability of one approach that should be employed more often than it currently is.

2.7 ACKNOWLEDGMENTS

This work was partially supported by the German Research Foundation (DFG) for financial within project C03 of SFB/Transregio 161. We would like to thank K-Marie Lahmer and Rickard Leandertz for their assistance in data collection, Johan Fagerlonn for sharing his original stimuli, and BrainProducts GmbH (Munich, Germany) for loaning us the necessary equipment for this study.

3

BRAIN RESPONSES TO SEMANTICALLY EQUIVALENT AUDITORY CUES

This chapter has been reproduced from an article published at CHI 2018: Glatz, C., Krupenia, S. S., Bülthoff, and H. H., Chuang, L. L. (2018). Use the right sound for the right job: Verbal commands and auditory icons for a task-management system favor different information processes in the brain. In Proceedings of the 2018 CHI Conference on Human Factors in Computing Systems. ACM, New York, NY, USA. DOI: http://dx.doi.org/10.1145/3173574.3174046

3.1 ABSTRACT

Design recommendations for notifications are typically based on user performance and subjective feedback. In comparison, there has been surprisingly little research on how designed notifications might be processed by the brain for the information they convey. The current study uses EEG/ERP methods to evaluate auditory notifications that were designed to cue long-distance truck drivers for task-management and driving conditions, particularly for automated driving scenarios. Two experiments separately evaluated naïve students and professional truck drivers for their behavioral and brain responses to auditory notifications, which were either auditory icons or verbal commands. Our EEG/ERP results suggest that verbal commands were more readily recognized by the brain as relevant targets, but that auditory icons were more likely to update contextual working memory. Both classes of notifications did not differ on behavioral measures. This suggests that auditory icons ought to be employed for communicating contextual information and verbal commands, for urgent requests.

3.2 INTRODUCTION

Auditory notifications are used extensively by in-vehicle interfaces to inform the user of important events. Whilst parking, for example, decreasing beep intervals could communicate the nearing distance between a driver and an obstacle. This raises the question: How should auditory notifications be designed? Should they be verbal notifications that communicate instructions explicitly or should they be recognizable auditory icons that denote a critical scenario?

Previous researchers have generally agreed on the essential design guidelines for auditory notifications (Nees and Walker, 2011). Auditory notifications need to be: (1) easily detectable (Graham, 1999; Mynatt, 1994; Edworthy, 1994), (2) readily discriminable against background noise (Nykänen, 2008; Liljedahl and Fagerlönn, 2010), (3) capture attention (Ho and Spence, 2005, 2006), and, after all these requirements have been fulfilled (4) easily interpretable (Mynatt, 1994; Edworthy and Hellier, 2006; Ulfvengren, 2003). However, there are many ways in which auditory notifications can be designed to comply with these criteria. Preference between different designs is often determined by user studies that evaluate performance or subjective feedback. Unfortunately, such measures can often contradict from one study to another or fail to discriminate between different designs.

Verbal commands and auditory icons represent two general classes of auditory notifications that are commonly employed. They are favored over synthetic sounds (Leung et al., 1997; Dingler et al., 2008) because they are based on prior user experiences. Hence, they are easily learned in novel use settings for their intended meanings. Nonetheless, no clear consensus has been established for preferring either verbal commands or auditory icons.

Besides performance and subjective measurements, brain responses to notifications can also serve as a way to evaluate auditory notifications. Surprisingly, this approach is rarely employed (although, see Lee et al., 2014). In particular, electroencephalography (EEG) reveals how the brain processes information – namely, the extent to which notifications capture attention and are interpreted – regardless of how this eventually influences observable performance or subjective feedback. In other words, EEG provides a higher functional resolution of the processes that take place between stimulus presentation and the elicited response. Therefore, we can evaluate auditory notifications based on how user brains respond to them, and not merely on how users respond to them. The understanding that we can gain from the additional information provided by EEG can contribute towards the design of more appropriate and effective notifications that are easy to use.

The increasing reliance on automation to perform tasks is transforming the role of notifications. While notifications used to be prized for their effectiveness as a 'call to action', they are increasingly used to update and inform users on the current situation, or to assist users in supervising automation. This trend is especially prevalent in the context of automated vehicles, whereby auditory notifications have been specially designed to update drivers of automated trucks on prevailing road conditions and to remind them to supervise logistical tasks. Arguably, such notifications might not be readily evaluated by behavioral responses alone, but by how they are processed by the brain for conveyed information.

In the current work, we answer the following four research questions (RQ1-RQ4) using EEG analysis. RQ1: Are auditory icons or verbal commands more easily detected? RQ2: Are auditory icons or verbal commands more discriminable? RQ3: Do auditory icons or verbal commands capture more attention? RQ4: Do auditory icons or verbal commands result in more context-updating of what has to be done next?

To summarize, this paper makes the following contributions:

1. EEG demonstrates that both types of notifications are equally detectable and orient attention to the same extent or rather more in an applied context.

2. By using EEG, we are able to show that verbal commands are discriminated more easily than auditory icons, but that auditory icons are more likely to update contextual working memory.

3. Evaluating brain responses reveals that auditory notifications should follow a purpose-oriented design. That is, verbal commands seem more suitable for urgent requests while auditory icons should be used to communicate less pressing contextual information. Unlike behavioral measures, EEG gives insight in different stages of processing a notification. Notably, these insights are obtained in a passive and unobtrusive way and cannot be gained through behavioral results only.

4. The results of this study demonstrate that laboratory findings of auditory EEG studies can be extended to more realistic environments, such as driving simulators. Qualitative comparisons suggest that results scale with experience and that the effect of notifications on brain responses increase with increasing relevance.

3.3.1 *Challenges for Designing In-vehicle Auditory Displays*

Sound design is a challenging task not only technically but even more so re-garding human perception. Human's auditory perception is influenced by a variety of factors such as emotions, memories, cognition, environment, previous experience, and the ability to understand speech (O'Callaghan, 2009). In the design of auditory displays, sounds that communicate information to the user, factors like these are relied upon (McGookin and Brewster, 2004). When not using words to communicate information but sounds, we talk about sonification (Walker and Nees, 2011). In this regard, sound designers have to consider the important aspect of making sounds recognizable and identifiable for what they were designed. At the same time, auditory notifications need to convey the level of urgency without annoying the user (Marshall et al., 2007). Especially for in-vehicle notifications, auditory displays whose meaning is ambiguous can have a negative influence (i.e. higher perceived workload, slower response times; Wiese and Lee, 2007). Hence auditory notifications should be evaluated, for example, based on their detectability, their position of presentation, their identifiability, and their conveyed meaning (Tuuri et al., 2007).

To optimize the design of auditory notifications, a two-stage pipeline has been proposed (Liljedahl and Fagerlönn, 2010). At an early first stage, designers consult with the user audience about the designed sound which feeds back into the design process. The second stage is an evaluation of the auditory display in a greater auditory context, simulating the use of the sound in its intended environment. While this is a step in the right direction for optimizing the design of effective auditory notifications, it relies on subjective user feedback, as well as performance measures, and does not reveal the actual influence auditory notifications have on the processing by the human brain.

3.3.2 *Auditory Notifications*

Auditory notifications can be categorized into two main groups, speech and non-speech. We differentiate non-speech sounds into auditory icons (representative sounds) and earcons (abstract synthesized sounds; Graham, 1999; Keller and Stevens, 2004). Unlike earcons, verbal commands and auditory icons are readily recognizable and do not require learning, which typically translates to faster responses (McKeown, 2005).

Verbal Commands

We often rely on speech to communicate our intentions to each another. Once proficiency is acquired in a given language, verbal commands can be relied on to communicate complex messages that can be unambiguously interpreted (Leung et al., 1997; Dingler et al., 2008). Thus, it is natural for humans to prefer speech notifications (McKeown, 2005).

Nonetheless, verbal notifications face the risk of being masked by or confused with real conversations (Oh and Lutfi, 1999). To circumvent this problem, contrivances could be introduced to make verbal notifications less human-like and more discriminable from real speech. This could be achieved by manipulating the pitch or other spectral properties of verbal commands. Pilots were found to discriminate easily between natural speech and synthesized speech (Simpson and Marchionda-Frost, 1984). Thus, presenting notifications in synthesized speech could prevent verbal commands from being confused with real conversations in our environment.

Another shortcoming of speech is that it is harder to spatially localize than other sounds, presumably because of its smaller bandwidth (Tran et al., 2000). However, this can easily be compensated by being interpreted unambiguously when it is used appropriately. Verbal commands can present unambiguous spatial information through their semantic context. For instance, presenting the word '*front*' from a front speaker results in fast response times to potential head-on collisions (Ho and Spence, 2005). However, if not used appropriately, speech can attract attention to the extent that it could interfere with other tasks such as driving (Spence and Read, 2003).

Auditory Icons

Auditory icons are sounds that represent real world events. These are sounds with stereotypical associations with the object or event/action that created the sound. For example, the sound of a car horn could indicate a safety critical situation that requires immediate attention and action. Being familiar sounds, auditory icons are easily learned for their intended function.

An advantage of auditory icons is that they are not easily masked by background speech (Leung et al., 1997; Strayer and Johnston, 2001; Ho and Spence, 2006). For instance, auditory icons are unlikely to be confused with a radio jockey's monologues. Nonetheless, auditory icons, like skidding tires or the car horn, can still be confused with real environmental occurrences. Also, auditory icons have been shown to be more likely in generating false alarms than abstract notifications (Gray, 2011). This is most likely due to the fact that humans might have overlearned certain cues (e.g., car horn). Given that background experiences are likely to be different across different users, auditory icons might be challenging to calibrate for their conveyed urgency.

Auditory icons are susceptible to misinterpretation because a single sound can represent more than one meaning (Graham, 1999; Mynatt, 1994). Depending on previous experience and the use-context, auditory icons can be recognized as an object (that generates the sound) or as the action that generated the sound (Mynatt, 1994; Gaver, 1989; McKeown, 2005). For example, the sound of screeching tires can be interpreted either as a proximal collision vehicle or as a command for braking. In complex operations, auditory icons might not be the appropriate notification (Graham et al., 1995; Haas and Schmidt, 1995). However, Belz et al. (1999) used skidding tires and a car horn honk in highly safety critical situations, namely to signal impending collisions. The successful use of auditory icons in this case might be due to the straight-forward association of meaning. In establishing guidelines for designing auditory icons, Mynatt (1994) suggests that auditory icons usability is highly affected by their identifiability. Nonetheless, the recognition accuracy for auditory icons, but not response times, can be significantly improved if users are aware of the icons' design (Lucas, 1994). This suggests that auditory icons should use mappings that do not have multiple interpretations and can be easily associated with the events they are representing.

Verbal Commands versus Auditory Icons

Auditory notifications can be evaluated on two different performance measures. First, effective notifications are believed to elicit faster reaction times. Second, effective notifications are more accurately detected from background noise and discriminated from other notifications. Previous research has compared the effectiveness of verbal commands and auditory icons across different scenarios and has generally found mixed support for either sound type.

To begin, while some studies have found faster responses for verbal notification (Lucas, 1994; Ho and Spence, 2005; Fagerlönn et al., 2015), others found faster responses for auditory icons (Graham, 1999; Saygin et al., 2005). Other studies have found a response time preference for neither auditory icons nor verbal notifications (McKeown, 2005; Ulfvengren, 2003). Overall, this would suggest that the processing for both sounds is not different. Some reasons for these mixed findings could be the context in which they were presented as well as the type of task participants were asked to do. Some studies, for example, took place in a simulated driving context and asked participants to avoid collisions by breaking while others required participants to match a presented sound to a description of the object/action generating this sound on a desktop PC.

Similarly, measures for response accuracy also provide mixed support for either verbal commands or auditory icons. Cummings et al. (2006) found that people were more accurate at matching auditory icons to visual context than words. This could be due to a trade-off between response times and discrimination sensitivity since reaction times were faster for nouns than auditory icons in this study. In contrast, Saygin et al. (2003); Orgs et al. (2006) found higher accuracy when cuing with verbal commands than with auditory icons. Leung et al. (1997), on the other hand, found no difference in accuracy for verbal commands compared to auditory icons. These different findings could be due to the different experimental tasks participants were asked to complete. On the other hand it could also depend on the type of auditory notifications employed. If time was not stressed, participants could respond when they had fully evaluated the auditory stimulus, even if the auditory icon was not readily interpretable at first.

Given that there will always be background noise, including meaningful sounds; it is worthwhile to use prominent and highly discriminable auditory notifications. Previous research has shown that distractor sounds have a larger negative impact on masking auditory icons than verbal commands (Saygin et al., 2005). That is, distractors were more likely to interrupt the processing of auditory icons than verbal commands. This suggests that non-verbal processing is more likely to be affected by increases in processing workload, often present in stressful and urgent situations.

3.3.3 Event-Related EEG Potentials

Brain responses can be used to evaluate auditory notifications, especially in terms of how they are processed for information. One prominent EEG measure is the event-related potential (ERP) that represents brain activity that proceeds from the presentation of a given stimuli. An ERP is an average waveform of negative and positive voltage deflections, which can be functionally related to different stages of information processing of the presented stimuli (Luck, 2005). Auditory stimuli characteristically elicit a series of evoked potentials, namely

N1, P2, P3a, and P3b components which are associated with how the presented sound is processed (Crowley and Colrain, 2004; Picton, 2011; Kraus and Nicol, 2009). Thus, amplitudes of these potentials can provide insight into how the brain processes a given sound.

N1

The N1 is the first negative deflection of the ERP waveform (e.g., Luck, 2005; Picton, 2011; Woods, 1995). It reflects early involuntary sensory processing and is highly sensitive to a sound's physical properties (e.g., pitch, intensity; Luck, 2005; Steinschneider and Dunn, 2002). It is a reliable indicator to the perceptual detection of a sound's presentation (Winkler et al., 2013; Kotz, 2014; Picton, 2011).

P2

The P2 is the second positive deflection, prominent at frontal and central electrode sites. It is often reported as part of the N1-P2 complex, also called 'vertex-potential' (Crowley and Colrain, 2004; Picton, 2011). In the current context, we interpret the P2 component as a measure for discriminability. It is elicited by attended and not attended stimuli (Crowley and Colrain, 2004; Picton, 2014). This means that it is evoked involuntarily by both, target and non-target stimuli of an oddball paradigm. The difference between the P2 of target and non-target stimuli is that the P2 amplitude is larger for stimuli containing target features (Luck, 2005). This fits the theory of P2 reflecting object discrimination (Novak et al., 1992; García-Larrea et al., 1992). P2 amplitudes are larger after one learns to discriminate a stimulus from other target stimulus (Alain, 2007). In this respect, a larger P2 amplitude reflects target identification and only subsequently a P3 is elicited. Once this classification has taken place, the resolved information can be transmitted to higher cortical areas to be evaluated further (Picton, 2011). If the stimulus being processed does not contain target features, no P3 is elicited and, hence, further cognitive evaluation of the stimulus is stopped.

P3a

The P3a refers to the third positive peak that is observed at frontal areas. It is evoked by unexpected stimuli regardless and decreases in amplitude to a surprising stimulus if it is presented repeatedly (Riggins and Polich, 2002). It is sometimes referred to as the novelty P3 and is believed to reflect an automatic orienting response to interesting information (Lee et al., 2014; Escera et al., 2000; Polich, 2007).

P3b

The P3b refers to the third positive peak that is observed at centro-parietal regions. It is sensitive to the presentation of task-relevant stimuli, especially those that occur infrequently (Duncan-Johnson and Donchin, 1977; Picton and Hillyard, 1974). In the current context, P3b is treated as a measure for context-updating, a process that underlies how we update our situational understanding when unexpected events occur (Donchin and Coles, 1988; Fabiani et al., 1986; Polich, 2007). When an interesting stimulus is recognized, which is different from the standard background, our brains update our mental representations of the environment. This updating process is reflected by larger P3b amplitudes for target stimuli than standard stimuli. Furthermore, P3b amplitudes are also treated as an index for working memory load, whereby larger P3bs are associated with less mental effort (Brouwer et al., 2012).

To date, little research has been conducted to investigate how verbal commands might be processed differently by the brain, compared to auditory icons. Behavioral studies have mixed results given that they tend to differ depending on the experimental task and context. The notifications that are used here were designed for an auditory display of a highly automated truck environment (Krupenia et al., 2014), for which verbal commands have been claimed to deliver faster responses than auditory icons (Fagerlönn et al., 2015). However, the underlying reason for this has been unclear. The current study was designed to look specifically at how the brain might process these auditory notifications differently, depending on whether they were verbal commands or equivalent auditory icons. For this purpose, we measured the EEG activity of naïve participants (Experiment 1) and professional truck drivers (Experiment 2). The current EEG dataset has been previously analyzed for differences between the two participant groups and have shown that both groups respond to these notifications similarly as a whole (Chuang et al., 2017). While professional truck drivers responded slower in general, this was not due to fundamental differences in brain responses to the auditory notifications. In fact, the current analyses show similar EEG/ERP waveforms between the two participant groups. In contrast to previous work, this current work focuses specifically on how verbal commands and auditory icons are processed differently in the brain. Although both types of auditory notifications produce similar brain responses, significant differences exist in specific ERP components, which suggests that they should be employed for different purposes.

3.4 STUDY METHODS

This study compared auditory icons to verbal commands in a controlled experimental laboratory environment. It was a within-subject design that used separable EEG/ERP components to evaluate these notifications for how well they were detected (N1), discriminated for against other sounds (P2), captured attention (P3a), and updated contextual working memory (P3b). The whole experiment lasted 2.5 hours, which included training, preparation time, and debriefing. The experimental procedure was approved by the Ethics Council at the University Hospital Tübingen.

Two experiments comprised this study. Experiment 1 was performed on university students (N=15; mean age=26.1 ± 4.0 years; 9 males) and a follow-up experiment 2, on professional truck drivers (N=15; mean age= 41.4 ± 12.1 years; 13 males). Experiment 1 was conducted in a psychophysical laboratory setting and experiment 2, in a high perceptual fidelity fixed-based truck simulator.

3.4.1 *Participants*

All participants reported no known hearing deficits, normal (or corrected-to-normal) vision, and no history of neurological problems. They provided signed consent to written instructions, and were remunerated for their voluntary participation.

3.4.2 *Stimuli and apparatus*

Auditory notifications

The auditory notifications (duration: 500 ms) were adapted from target sounds that were originally designed for the in-vehicle interface of an autonomous truck cabin (Fagerlönn et al., 2015; Krupenia et al., 2014). There were 12 notifications in total, 6 verbal commands and 6 auditory icons that were complements of each other. They were designed to remind truck drivers to perform certain tasks and of driving conditions at the appropriate times. These verbal commands (*auditory icons*) were "system" (*synthetic tone*), "convoy" (*train whistle*), "driver" (*human whistle*), "weather" (*raindrop*), "road" (*ground rumbling*), and "traffic" (*car horn*). Verbal commands were in Swedish, which was the mother tongue of the professional drivers in Experiment 2 but not the student volunteers of Experiment 1. However, the student volunteers were extensively briefed on the auditory stimuli and practiced discriminating them until they were 80% accurate, prior to testing.

Ninety distractor sounds were created. Each distractor was a simultaneous presentation of 2 verbal commands and 2 auditory icons, played in reverse and their loudness adjusted to be comparable to the notification targets.

Experiment 1: Psychophysics laboratory

The experimental laboratory was a dark room which was insulated against external sounds. The visualization was presented on a desktop screen (ViewPixx Screen, 60.5 x 36.3 resolution; 120 Hz) at a fixed distance of 45 cm from the participant who was in a chin-rest. The experiment was controlled with customized software (MATLAB 8.2.0.701, R2013b) and Psychophysics Toolbox 3.0.12 (Brainard, 1997; Pelli, 1997; Kleiner et al., 2007). An ASIO 2.0 compatible sound card was used to control sound presentation (SoundBlaster ZxR; Creative Labs). The auditory stimuli were presented via stereo speakers, one paced on the left the other on the right side of the desktop display.

Experiment 2: High fidelity truck simulator

Professional truck drivers sat in a driving simulator that consisted of a realistic truck cabin. This contained a steering wheel, dashboard with instruments, and a pneumatic seat. The visualization consisted of an automated drive on the highway from Linköping to Norrköping with minimal traffic. The highway had two lanes, one for each antagonistic traffic direction. A three wall display (approx. 150 deg field-of-view) presented the frontal visualization (450 cm distance to head). Two vertically-aligned displays were attached to the outside of the cabin, to simulate side mirrors for displaying the rear traffic scene. Using OpenDrive8 road network files (xodr) and an additional file describing the landscape (xml), a customized graphical engine (i.e., VISIR) rendered the presented visualization. Buttons located on the left and right on the steering wheel collected the participants' behavioral responses. An ASIO 2.0 compatible sound card (RME HDSP 9632; RME Intelligent Audio Solutions) controlled the presentation of auditory notifications. A 5.1 surround sound system, installed in the truck cabin, was used to display the sounds.

3.4.3 Task and Procedure

During testing all participants observed an automated driving scene that was visually presented. They had to respond to auditory notifications whenever one was presented, with a button press using either their left or right index fingers. Six notifications (i.e., 3 complementary pairs of verbal commands and auditory icons) were pre-assigned to a left index-finger press and the remaining six, to a right index-finger press. In other words, each button corresponded to three events, which were represented by a verbal command as well as an auditory icon. Button press assignment was randomized across participants. Prior to testing, all participants practiced until they were able to achieve accuracy levels of at least 80% in this task.

Approximately 980 sounds were presented throughout the experiment testing. All sound presentations were separated by a time-interval randomly selected from a uniform range of $2300 - 2700$ ms. Twenty percent of these were target notifications and eighty percent were distractors. No button presses were necessary when distractors were presented. Target notifications were evenly divided for verbal commands and auditory icons. Participants had 2000 ms to respond after each auditory notification was presented.

3.4.4 EEG recording

The EEG was recorded using 59 active electrodes mounted to the scalp using an elastic cap according to the international 10-20 system (ActiCap System, Brain Products GmbH, Munich, Germany). Four additional electrodes were used to record the vertical and horizontal electrooculogram from the right and left canthi as well as above and below the left eye. All signals were recorded with the online reference FCz and AFz as the ground. EEG signals were digitized with a sampling rate of 1000 Hz. Electrode gel was applied to each electrode to ensure an impedance below 20 kΩ. A parallel port connection between recording PC and experimental PC synchronized the EEG recording with the experimental events, such as the sound onset and button press.

To analyze the EEG data, MATLAB (8.2.0.701, R2013b) and EEGLAB v.14.0.0[1], an open source software to analyze electrophysiological data, was used (Delorme and Makeig, 2004). Before analyzing the ERP to the auditory stimuli, the data was preprocessed for every subject according to the following steps. To reduce the computational costs, the data recorded at 1000Hz was downsampled to 250 Hz. To remove any slow drifts, a high-pass filter (cut-off = 0.5 Hz) was applied subsequently to the data. Using CleanLine, a plugin in EEGLAB, 50 Hz electrical noise picked up from the environment when recording electrical brain activity was removed. Then, bad channels, e.g. channels with flat lines, are removed using artifact subspace reconstruction. Following these cleaning steps, the data is re-referenced offline to the common average reference and then submitted to the Adaptive Mixture ICA (AMICA; Delorme et al., 2012). This algorithm decomposes the electrical activity recorded at sensor level (electrodes) into source-resolved activity, also called independent components (ICs). These ICs were subject to equivalent dipole estimation. A MNI Boundary Element Method head model was used to fit an equivalent dipole to the ICs (Piazza et al., 2016). IC dipoles with location outside the brain, as well as, ICs with a residual variance larger than 15% were excluded. Next, the ICs of all participants were grouped into 30 clusters using k-means based on their power spectrum. These clusters were then inspected for non-cortical electrical activity such as eye-related activity, muscle-related activity, line noise, and unresolved components. Clusters, containing such non-cortical activity, were determined based on their power spectrum, their scalp topography, and their dipole location in a volumetric brain model. These non-cortical activity clusters, present across the group of participants, were removed from the EEG data (for examples, see Chuang et al., 2017). Finally, this EEG data for cortical activity was backprojected to the sensor level, and analyzed for potential differences between verbal commands and auditory icons.

ERPs were computed for each participant, and for every electrode, by extracting an epoch of EEG activity around the notification presentation. The presentation onset of the notifications was the trigger event for an epoch that consisted of 500 ms of baseline activity pre-trigger and 1000 ms of brain response post-trigger. All epochs that belonged to either verbal commands or auditory icons were mean-averaged for each electrode. We further grouped the frontal and pari-

---

[1] https://sccn.ucsd.edu/eeglab/

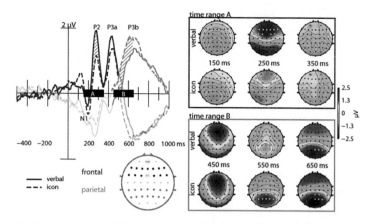

Figure 6: Experiment 1's ERP responses (left) with scalp topography plots (right) of statistically significant differences across time and electrodes respectively. Left: ERP waveforms are averaged across the frontal (pink) and parietal (green) electrodes and deflections are labeled for N1, P2, P3a, and P3b. The shaded areas between the two waveforms indicate time-regions that are significantly different. Right: The scalp topographies show the EEG activity to verbal commands and auditory icons at time-ranges A and B. Electrodes that are significantly different are represented by white dots.

etal electrodes into two separate groups for visualization (see Figures 6 and 7, right). These group averaged waveforms depict distinct ERP components (i.e., N1, P2, P3a, and P3b) that serve as established neural correlates for perceptual and cognitive mechanisms. With regards to auditory information processing, they relate to detection (N1), discrimination (P2), attentional capture (P3a), and context-updating (P3b).

We performed mass-univariate analysis (MUA) to statistically evaluate EEG differences between verbal commands and auditory icons (Groppe et al., 2011). Simply put, this method compares the two conditions at every electrode and time-point and performs a t-test for significant differences. A false discovery rate procedure (i.e., FDR-BH; Benjamini and Hochberg, 1995) was applied to control for multiple comparisons.

3.5.1 Behavioral performance

Behavior performance was evaluated in terms of discrimination sensitivity and correct response times. All scores were submitted to a within-subjects t-test and the Bayes Factor (BF_{01}) was calculated for the likelihood of the null-hypothesis relative to the alternative-hypothesis. The behavioral data of one participant from Experiment 2 had to be excluded due to missing button presses.

Discrimination sensitivity (d') was computed for each participant as the difference between the z-score of correct recognition and false recognition. In Experiment 1, d' scores were not significantly different for verbal icons and auditory icons ($t_{14} = 0.17, p = 0.87, Cohen's\ d = 0.04$), respectively 3.82 ± 1.15 and 3.78 ± 1.10. The null-hypothesis was favored by a Bayes Factor of 3.7. In Experiment 2, d' scores were not significantly different for verbal icons and auditory icons ($t_{13} = 1.53, p = 0.15, Cohen's\ d = 0.41$), respectively 2.45 ± 1.44 and 2.17 ± 1.12. The null-hypothesis was marginally favored by a Bayes Factor of 1.4.

Response times were calculated for all correct responses that occurred within 2500 ms of notification onset, for each participant. In Experiment 1, participants were not significantly faster in responding to verbal commands compared to auditory icons ($t_{14} = 0.68, p = 0.51, Cohen's\ d = 0.18$), respectively 1070 ± 131 and 1094 ± 182 ms. The null-hypothesis was favored by a Bayes Factor of 3.1. In Experiment 2, participants were not significantly faster in responding to verbal commands compared to auditory icons ($t_{13} = 0.46, p = 0.65, Cohen's\ d = 0.12$), respectively 1236 ± 101 and 1251 ± 156 ms. The null-hypothesis was favored by a Bayes Factor of 3.3.

To summarize, the current behavioral results do not argue in favor of either verbal commands or auditory icons.

3.5.2 EEG/ERP responses

The EEG/ERP activity elicited by verbal commands and auditory icons were similar in general morphology, latency, and scalp distribution in the anterior-posterior dimension for both Experiments 1 and 2 (Figs. 6 and 7). Statistically significant differences were revealed in the EEG/ERP activity generated by auditory icons and verbal commands in the frontal as well as parietal electrodes[2].

---

[2] Electrode labels are in italics to distinguish them from ERP component labels.

The frontal group of electrodes are: F_5, F_3, F_1, F_z, F_2, F_4, F_6, FC_5, FC_3, FC_1, FC_2, FC_4, FC_6. The parietal group of electrodes are: P_5, P_3, P_1, P_z, P_2, P_4, P_6, CP_5, CP_3, CP_1, CP_z, CP_2, CP_4, CP_6.

In Experiment 1, student participants showed significant differences in their EEG/ERP responses to these notifications, even though the verbal commands were not presented in their native language nor did they have contextual meaning. Specifically, the amplitude of the P2 component (236 − 304 ms; frontal electrodes) was significantly larger for verbal commands than for auditory icons. This suggests that verbal commands were more discriminable than auditory icons from presented sounds. The P3b amplitude (512 − 640 ms; parietal electrodes) was significantly larger for auditory icons than for verbal commands. This suggests that auditory icons induced more context updating than verbal commands did.

In Experiment 2, the professional truck drivers generated similar results as the naïve participants of Experiment 1. Similarly, verbal commands generated larger P2 component deflections (212 − 352 ms; frontal electrodes) than auditory icons, and auditory icons generated larger P3b deflections (412 − 624 ms) than verbal commands. However, EEG/ERP activity in the frontal electrodes revealed that auditory icons generated larger N1 deflections (160 − 212 ms) than verbal commands, and that verbal commands generated larger P3a deflections (352 − 468 ms) than auditory icons. Differences in N1 deflections suggest that auditory icons were more likely to be detected against the general auditory background. Differences in P3a deflections suggest that verbal commands were more likely than auditory icons to capture observer attention.

3.6 DISCUSSION

The design of auditory displays faces the challenges of in-cooperating human perception to make notifications effective for their designed purpose (see Sect. 3.2 and 3.3). Behavioral performance measures are limited in discriminating between notifications for the various purposes that they might be designed for. Some notifications might be designed for the purposes of being highly detectable while others might be designed to communicate a given context or scenario. Performance measures, e.g., response times or accuracy, do not discriminate for how the brain processes notifications for information.

To summarize, we evaluated verbal commands and auditory icons that were especially designed for in-vehicle information displays related to aspects of task

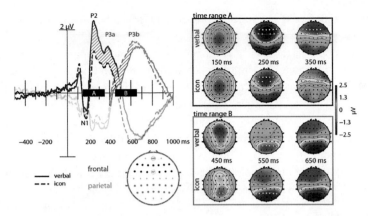

Figure 7: Experiment 2's ERP responses (left) with scalp topography plots (right) of statistically significant differences across time and electrodes respectively. Left: ERP waveforms are averaged across the frontal (pink) and parietal (green) electrodes and deflections are labeled for N1, P2, P3a, and P3b. The shaded areas between the two waveforms indicate time-regions that are significantly different. Right: The scalp topographies show the EEG activity to verbal commands and auditory icons at time-ranges A and B. Electrodes that are significantly different are represented by white dots.

management and context-updating in highly automated trucks (Fagerlönn et al., 2015; Krupenia et al., 2014). We were motivated to do so according to the guidelines for auditory displays (Nees and Walker, 2011), which states that auditory displays ought to be highly detectable in terms of their physical properties (N1), support learned discrimination from other target notifications (P2), have the potential for capturing attention (P3a), and communicate for its intended purpose (e.g., updating contextual working memory; P3b). Experiment 1 tested naïve students, as is often the case during design and prototyping phases, and Experiment 2 tested professional truck drivers, as is often the case when performing an evaluation and validation. Both types of notifications were effective and did not significantly differ in terms of response times or accuracy. However, verbal commands are more easily discriminable from other target notifications at an early perceptual stage (i.e., larger P2 component), and auditory icons are more likely to update contextual working memory (i.e., larger P3b). This is consistently true regardless of testing environment or participant groups. In fact, discriminable and significant trends in EEG/ERP waveforms are amplified in professional truck drivers. Professional truck drivers also show a neural preference for verbal commands with regards to their detection (N1) and attentional capture properties (P3a), which is presumably driven by a familiarity with the given language.

Therefore, we advocate the use of both verbal commands and auditory icons. However, they should be employed according to the job that they are intended for. We suggest that verbal commands should be employed in critical situations that require immediate action while auditory icons seem more appropriate to notify the user of non-urgent environmental updates. In the context of highly automated vehicles, verbal commands should be used for time-critical situations that cannot afford ambiguity, such as *'low fuel'*, while auditory icons might be better employed in indicating that driving conditions are changing, such as the sound of a thunderstorm for inclement weather.

3.6.1 *Justification for the Current Interpretation*

Our recommendations for using verbal commands and auditory icons for different purposes are based on the following reasons, based on the EEG/ERP responses to these auditory notifications. We do not address the performance results given that they are shown to be equally effective in terms of discriminability and correct response times.

To begin, we do not emphasize the detectability of either notification types (in terms of their physical properties). This is because the timing of N1s were identical in both Experiments 1 and 2. Interestingly, the amplitude of N1 was larger for auditory icons in Experiment 2. We believe that this reflects the larger variability of the spectral properties of auditory icons relative to verbal commands, which renders it more detectable against a richer (i.e., noisier) background. While this could be treated in favor of auditory icons, we believe that the consistently larger amplitudes in P2 components for learned notification discriminability across both experiments compensates for this minor advantage.

Next, the frontal P2 component is larger for verbal commands than for auditory icons. P2 is believed to reflect learned object discrimination (Crowley and Colrain, 2004; Novak et al., 1992; García-Larrea et al., 1992). In this regard, P2's amplitude indicates the efficiency in recognizing the associated notification as a discriminable target, relative to other target notifications. Trained discriminability is known to have an effect on P2 amplitude. For example, musicians who are trained to discriminate sounds for pitch and timbre generate larger P2s than non-musicians, especially for musical sounds (Shahin et al., 2003). In a similar fashion, most of us are highly trained to discriminate between different verbal sounds for their intended meanings and associations. This reflects the natural advantage rendered by the use of verbal commands over auditory icons. The current findings suggest that even if certain auditory icons are determined by sound designers as being highly discriminable and recognizable as targets, they should be matched to the standards of verbal commands, which is quantitatively measurable in terms of P2 component amplitude.

Thirdly, we believe that verbal commands capture attention more readily than auditory icons. P3a amplitudes are indicative of an involuntary orienting response to surprising and novel events (Polich, 2007). While the larger P3a amplitude for verbal commands was not significant in Experiment 1 (Figure 6), it was for the participants of Experiment 2. We believe that this was because the professional truck drivers understood the verbal commands and their operational implications more readily, which increased the potential of verbal commands's in capturing attention (Figure 7).

Last, but not least, the P3b component reflects the updating of one's mental representation of relevant information (Donchin and Coles, 1988). P3b amplitudes are larger when a task-relevant event occurs that is different from one's expectation. For this reason, it is believed to underlie context-updating (Polich, 2007). Related to this, P3b amplitudes have also been used to evaluate for working memory load or mental workload (Brouwer et al., 2012). Larger P3b are associated with low mental workload and smaller P3b, with high mental workload. In the current context, this would suggest that auditory icons are more memorable and result in stronger context-updating than verbal commands, in a way that requires significantly less mental effort.

Auditory displays are designed to capture attention and to clearly communicate events. On the one hand, notifications that are readily recognized as task-relevant targets and capture attention are necessary for urgent events. On the other hand, notifications to indicate changing circumstances are also required to assist in updating a user's situational awareness. The current EEG/ERP results indicate that verbal commands are more discriminable and better at capturing attention than auditory icons. This suggests that verbal commands should be used in critical situations where quick action is required. Previous research based on behavioral results are in agreement with our conclusion. For example, verbal notifications are claimed to be especially effective in stressful situations because speech is processed automatically (Graham, 1999). The current EEG/ERP results also suggest that auditory icons result in less effortful context-updating than verbal commands. Hence, auditory icons appear to be more suitable in communicating environmental circumstances that are less urgent. Edworthy and Hards (1999); Keller and Stevens (2004) have suggested using auditory icons as notifications that inform and advise about background events. Their recommendation based on behavioral results agrees with our current findings. It is worth noting that auditory icons are also believed to produce higher compliance levels than verbal commands, if they are understood in the first place (Edworthy, 1994). In conclusion, our results advocate that different types of notification should be designed in accordance to their intended purpose. This agrees with previous findings that, until now, have been mixed due to the imprecision of behavioral results in discriminating for how these notifications might be processed by the brain.

3.6.2 *Limitations of the Current Study*

One might argue that we only observe the changes in the ERPs because the neural origins of auditory icons and verbal commands are possibly oriented differently. This is a known limitation of ERP analysis, namely that it has limited spatial resolution for localizing brain regions that give rise to detected activity. However, this line of argument is unlikely. Previous work that relied on neuroimaging with better spatial resolution (i.e., fMRI) have demonstrated that words and object sounds involve the same neural region for processing information content (Dick et al., 2007). In addition, verbal commands and auditory icons produce equivalent scalp topographies (Cummings et al., 2006), which we have also observed in the current experimental paradigm. Therefore, it is unlikely that we observed the current results because verbal commands and auditory icons were not processed by identical neural regions.

The differences between the participant groups of Experiments 1 and 2 were intended to reflect the different stages of notification development, namely design and prototyping (Experiment 1) and validation (Experiment 2). Nonetheless, some differences between the participant groups might be of concern. In particular, the student volunteers in Experiment 1 were not proficient in the verbal commands. In spite of this, we note that their brain responses to verbal commands and auditory icons replicated in Experiment 2, which employed truck drivers. One reason is that the native language of Experiment 1's participants was highly similar to Experiment 2's participants. Another reason is that the current task focused on the participants' ability to recognize notifications and to respond appropriately, regardless of the extent to which they understood the implications of the notifications. Professional truck drivers who understood the language and the implications of the notifications showed stronger neural discriminations between verbal commands and auditory icons. Nonetheless, we can argue that the trends observed in Experiment 1 are likely to be generalizable trends while those in Experiment 2 are cultural and profession specific. The current study does not directly compare the two participant groups for their discrimination ability of the given sounds. Therefore, we do not make any inferences concerning either group's proficiency in discriminating the notifications from one another. The focus of this work is in evaluating how verbal commands and auditory icons are responded to at the level of information processing (i.e., brain responses).

3.7 CONCLUSION AND FUTURE WORK

Taken together, the current work contributes by showing that auditory notifications can be evaluated and functionally discriminated for how they are processed by the brain for information. This has implications for the operational context as well as design. Choices for which notifications to use for which purpose can be based not only in terms of response times and discrimination accuracy, which is not necessarily the operational objective, but in terms of how the notifications are: (1) detected against the auditory scene, (2) discriminated against other notification targets, (3) likely to capture attention, and (4) capable of updating contextual working memory.

To date, most studies have questioned whether verbal commands or auditory icons serve better as notifications, namely in terms of how well they elicit a speeded and accurate response. The current findings suggest that this question, while well-intentioned, is misplaced. Our results demonstrate that verbal commands and auditory icons have different qualities. While verbal commands are better discriminated against other notifications, auditory icons can update contextual working memory with less effort. Practically speaking, this suggests that verbal icons are ideally used for time-critical information where there is no leeway for ambiguity, e.g., collision warnings. Meanwhile, auditory icons are likely to be more effective in communicating contextual information, such as entry into a poorly maintained road section or changing weather conditions. In other words, verbal commands and auditory icons should be used as complementary (and not competing) notifications.

Previous research has recommended using auditory icons to notify users of environmental events (Edworthy and Hards, 1999; Keller and Stevens, 2004). More specifically, auditory icons have been suggested to enhance situational awareness (Adcock and Barrass, 2004; Kazem et al., 2003). For example, a walking sound can more effectively indicate a nearing pedestrian. In addition, auditory icons might be favored because it is believed that they can be processed in parallel to other auditory events (Adcock and Barrass, 2004). These findings so far converge with our current results and interpretation. Nonetheless, there are works that do not. For example, contrary to our current believe, that verbal commands capture attention, some work have shown that certain auditory icons (i.e., car horn) result in significantly faster response times (e.g., Graham, 1999). We might account for this by the fact that some auditory icons are overlearned to indicate danger. It should be noted that verbal processing is known to differ for different word classes (i.e., verbs, nouns; Pulvermüller, 1999; Szekely et al., 2005). The current study only uses nouns for verbal commands and, thus, future studies should verify whether verbal commands attract attention preferentially for all word classes, relative to auditory icons. In this work, we present EEG/ERP evidence that discriminates for how auditory icons and verbal commands are processed by the brain. Nonetheless, we do not doubt that nuances in how auditory notifications are engineered could ultimately render an auditory icon attention-grabbing and/or a verbal command more suited for communicating context. Our current results contribute by providing a starting point for understanding what type of sounds ought to be employed for which purposes, bearing in mind the brain's likely response to them.

The participants in Experiment 1 possessed neither a language proficiency for the verbal commands nor an expert understanding of the operational tasks that the notifications indicated. Therefore, the EEG/ERP differences (i.e., P2, P3b) found between verbal commands and auditory icons can be considered as general differences between the two notification classes. In contrast, Experiment 2 was performed on professional truck drivers in a highly realistic test environment. A comparison between the two experiments reveals that these differences in brain responses scale with realism and user proficiency. Thus, the current approach of evaluating notification designs on the basis of brain responses is robust, even when behavioral responses do not differ. Notifications that are first designed in sterile lab environments could also be evaluated for the EEG/ERP responses that they elicit. This would narrow down the candidates for deployment and validation in high fidelity simulation environments or field-testing.

Besides this, EEG/ERP methods could also be used to discriminate between different instantiation of the same target notification. One example would be to determine the preferability of semantically comparable verbal commands, such as *tank* or *fuel*.

To conclude, the current work suggests that verbal commands and auditory icons serve different purposes, at least from the standpoint of how they are processed by the brain. Thus, evaluations that directly compare them in terms of performance measures might not be appropriate. This might also explain the mixed evidence from previous studies in support of either auditory notifications. The growing accessibility of brain recording methods (i.e., EEG) mean that the current approach can be used to support finer functional discriminations for notifications and can be effectively deployed, even in challenging deployment scenarios such as high fidelity truck simulators.

3.8 ACKNOWLEDGMENTS

We thank the reviewers for their valuable feedback, which was helpful in revising the paper. We also thank K-Marie Lahmer and Rickard Leandertz for their assistance in data collection, and BrainProducts GmbH (Munich, Germany) for loaning us the necessary equipment for this study. This work was supported by the German Research Foundation through SFB/Transregio 161 projects, as well as by Scania CV AB, Sweden.

TEMPORAL DYNAMICS OF AUDITORY LOOMING CUES

This chapter has been reproduced from an article that was submitted for publication: Glatz, C. and Chuang, L. L. (2018). The time course of auditory looming cues in redirecting visuo-spatial attention.

4.1 ABSTRACT

By orienting attention, auditory cues can improve the discrimination of spatially congruent visual targets. Looming sounds that increase in intensity are processed preferentially by the brain. Thus, we investigated whether auditory looming cues can orient visuo-spatial attention more effectively than static and receding sounds. Specifically, different auditory cues could redirect attention away from a continuous central visuo-motor tracking task to peripheral visual targets that appeared occasionally. To investigate the time course of crossmodal cuing, Experiment 1 presented visual targets at different time-points across a 500 ms auditory cues presentation. No benefits were found for simultaneous audio-visual cue-target presentation. The largest crossmodal benefit occurred at early cue-target asynchrony onsets (i.e., CTOA=250 ms), regardless of auditory cue type, which diminished at CTOA=500 ms for static and receding cues. However, auditory looming cues showed a sustained cuing benefit at CTOA=500 ms. Experiment 2 showed that this late auditory looming cue benefit was independent of the cues intensity when the visual target appeared. Thus, we conclude that the sustained benefit throughout an auditory looming cues presentation is due to its increasing intensity profile. The neural basis for this benefit and its ecological implications are discussed.

4.2 INTRODUCTION

In the real world, we rarely perform a single task without being interrupted. For example, driving a car requires us to concentrate on the road ahead while remaining alert to events in our visual periphery, such as the sudden appearance of a jaywalking pedestrian. Cues can orient our attention to the spatial location of a congruent target that appears thereafter, resulting in a cuing benefit in terms of more accurate target discrimination and/or faster correct response times (Müller and Rabbitt, 1989; Posner et al., 1980; Posner, 1980; Eriksen and Hoffman, 1972; Egly and Homa, 1991; Henderson and Macquistan, 1993). Cuing benefits can be intramodal (i.e., visual cues and visual targets; Posner, 1978; Egeth and Yantis, 1997; Pashler, 1998) as well as crossmodal — for example, but not exclusive, to spatially congruent auditory cues and visual targets (Driver and Spence, 1998; Spence and Driver, 2004). It is well-established that auditory cues can orient attention so as to enhance processing of spatially congruent visual targets (McDonald et al., 2000; Lee and Spence, 2015; Ho and Spence, 2005). In this study, we investigated whether an auditory cue that increases with intensity over time (i.e., an auditory looming cue) induced cuing benefits to visuo-spatial attention, superior to a static auditory cue. We report that auditory looming cues are especially effective in redirecting spatial attention from a continuous central manual tracking task to an occasionally occurring peripheral discrimination task. Interestingly, we find that this benefit only manifests itself at later cue-target onset asynchronies (CTOA; i.e., 500 ms). This introduction focuses on the motivation to investigate auditory looming cues' influence on visual spatial attention.

Objects appear to approach us when they expand visually or when they get louder with time. Such objects, termed *looming*, are claimed to be especially salient because they could signal imminent threats. For example, visual looming objects induced involuntary fear and avoidance responses in mice (Yilmaz and Meister, 2013), rhesus monkeys (Schiff et al., 1962), and human infants (Ball and Tronick, 1971), which suggests reflexive mechanism to looming stimuli. The auditory equivalent, namely looming sounds with rising intensities, have also been associated with preferential processing and alerting responses. Looming sounds with rising intensities are often perceived as changing more compared to equivalent receding sounds with falling intensities (Neuhoff, 1998; Bach et al., 2008; Neuhoff, 2001). Also, looming sounds elicit larger skin conductance responses (Bach et al., 2009) and amygdala activity (Bach et al., 2008) than receding sounds. Finally, looming sounds are associated with greater activity in the auditory cortex, as well as in neural networks related to attention and spatial processing (Seifritz et al., 2002; Maier and Ghazanfar, 2007; Bach et al., 2015). Taken together, it is generally agreed that looming sounds are salient auditory stimuli that increase phasic alertness, presumably because they communicate approaching threats (Neuhoff, 2001).

The saliency of looming sounds could also influence visual perception. For example, static visual targets are perceived as larger or brighter than they really are, when accompanied by looming sounds (Sutherland et al., 2014). In a more realistic setting, drivers braked earlier if a potential head-on collision was accompanied with a looming sound, relative to a static auditory warning (Gray, 2011). More interestingly, auditory looming stimuli can induce excitation in visual cortical regions for low-level processing (Romei et al., 2009; Cappe et al., 2012). These interactions are often discussed in terms of multisensory integration (Cappe et al., 2012, 2009). Nonetheless, there is some evidence that looming sounds can also exert a preferential bias on visual spatial attention. When presented in only one ear, a looming sound can increase tilt discrimination sensitivity in the congruent visual hemifield relative to the opposing hemifield, for an object that is presented simultaneously (Leo et al., 2011). This raises the question: What is the role (if any) of a looming sound in reorienting visual spatial attention?

Salient events (e.g., abrupt onsets; Yantis and Jonides, 1990) can attract the rapid redeployment of spatial attention to themselves. For example, targets in a visual search array benefit from faster responses if they shared the same location as a preceding visual looming cue that increased in size, compared to if the cue decreased in size (Franconeri and Simons, 2003). Interestingly, the reorienting of attention by salient cues is extremely difficult to ignore (Cheal et al., 1991; Jonides, 1981; Giordano et al., 2009; Nakayama and Mackeben, 1989; Yantis and Jonides, 1990). Furthermore, this involuntary shift of attention can occur even when doing so impairs performance (Yeshurun and Carrasco, 1998; Yeshurun and Levy, 2003). Although salient events result in early spatial attention reorienting, this shift is transient and decays rapidly (Cheal et al., 1991; Müller and Rabbitt, 1989; Nakayama and Mackeben, 1989). In fact, the cuing benefit of transient attention reverses at long CTOAs, whereby RTs become longer at the previously attended location compared to unattended locations—an effect termed Inhibition of Return (IOR; Posner and Cohen, 1984; Posner et al., 1985) that is postulated to prevent re-inspection of previously explored locations and, hence, optimize visual search (Klein, 2000; Lupianez et al., 2006). Transient attention can be cued crossmodally, whereby auditory (Spence and Driver, 1994) and tactile (Spence and McGlone, 2001) cues can bias spatial attention toward spatially congruent regions of the visual field (see reviews Spence et al., 2004; Driver and Spence, 1998). Given the established saliency of looming sounds, it is highly plausible that looming sounds can serve as transient cues, which could have given rise to the effects reported by Leo and colleagues (Leo et al., 2011).

To sum up, looming objects are salient events that are preferentially processed by the perceptual system. Multisensory research suggests that looming sounds can enhance the perception of visual targets by exciting activity in the visual cortex, presumably because they signal relevance or threat (Romei et al., 2009). In agreement, peripherally presented looming sounds have been reported to enhance visual processing in the corresponding hemifield, relative to the opposite hemifield (Leo et al., 2011). Given these findings, it is surprising that auditory looming cues are rarely considered in terms of their ability to capture and reorient visual spatial attention. The current study investigates whether auditory looming cues might induce a crossmodal cuing benefit for spatial orienting that is different from equivalent auditory cues (i.e., static and receding intensity sounds). If salient looming sounds exert a strong influence on reorienting transient attention, cuing benefits ought to be larger at short CTOAs (i.e., 250 ms) and be smaller at long CTOAs (i.e., 500 ms) compared to other equivalent cues. Besides this, simultaneously presented auditory looming sounds at 0 ms CTOAs could enhance visual processing, independent of cuing benefits, given that they are known to influence early visual processing (Romei et al., 2009; Cappe et al., 2012). The results of this study indicate that neither predictions are true.

Experiment 1 varied the CTOAs between different auditory cues and the visual target of a peripheral tilt-discrimination task. It demonstrated large and comparable cuing benefits at 250 ms regardless of auditory cues, which greatly decreased with larger CTOAs of 500 ms for static and receding cues but less so for looming cues. Given that this could have resulted from the intensity differences between the auditory cues at 500 ms, Experiment 2 manipulated the final intensity levels of static and looming cues and compared their cuing benefits at a fixed CTOA of 500 ms. Experiment 2 verified that the cuing benefit of auditory looming cues, unlike auditory static cues, was independent of their intensity when the visual target appeared. Two key aspects set the current study apart from previous research. Unlike most studies on spatial orienting, we employed a dual-task paradigm that required participants to perform a central manual tracking task at all times. In other words, a sustained diversion of spatial attention away from the central location comes at a cost, which is presumably larger than if participants were merely requested to maintain central fixation. These differences between the current study and previous studies on looming sounds are discussed to appreciate the role of looming sounds on redirecting visuo-spatial attention.

4.3 EXPERIMENT 1: DO AUDITORY LOOMING SOUNDS ENHANCE PERIPH-ERAL TILT-DISCRIMINATION PERFORMANCE ACROSS ITS PRESENTED DURATION?

4.3.1 *Results and Discussion*

Performance in the peripheral tilt-discrimination task was operationalized in terms of the time that a participant took to respond correctly from the time of target appearance (RTs). To compensate for positive skews in RT measures (Ratcliff, 1993), medians RTs were calculated for each experimental condition. This data is presented in Figure 8. There is a general pattern of cuing benefits, irrespective of auditory cue types, that peaks for visual targets that appear 250 ms after the onset of the auditory cue and diminish for those that appear 500 ms after cue onset. Interestingly, there appears to be no benefit for the discrimination of visual targets that appear simultaneously with the auditory cues.

The median RTs were submitted to repeated-measure ANOVAs (JASP Team, 2018, see Supplementary Material) for the factors of *Auditory Cue* (none, looming, receding, static) and *CTOA* (0, 250, 500 ms). There was significant interaction between the factors of *Auditory Cue* and *CTOA* ($F(6, 84) = 13.383, p < 0.001, \omega^2 = 0.449$) as well as for both main effects (*Auditory Cue*: $F(3, 42) = 20.553, p < 0.001, \omega^2 = 0.560$; *CTOA*: $F(1.374, 19.242) = 9.155, p = 0.004, \omega^2 = 0.345$). To interpret the interaction, we performed separate one-way ANOVAs for the factor of *CTOA* for each auditory cue condition. With the exception of the 'none' condition, all conditions returned a significant main effect for *CTOA*. For auditory static cues, significantly faster RTs were found at a CTOA of 250 ms, compared to 0 ms ($t(14) = 4.746, p_{bonf} < 0.001, d = 1.226$) and 500 ms ($t(14) = 2.798, p_{bonf} = 0.028, d = 0.722$). Auditory receding cues showed a similar pattern of faster RTs at a CTOA of 250 ms, compared to 0 ms ($t(14) = 5.152, p_{bonf} < 0.001, d = 1.330$) and 500 ms ($t(14) = 4.354, p_{bonf} < 0.001, d = 1.124$). The RTs between CTOAs of 0 ms and 500 ms neither differed for auditory static cues ($t(14) = 1.949, p_{bonf} = 0.184, d = 0.503$) nor auditory receding cues ($t(14) = 0.798, p_{bonf} = 1.000, d = 0.206$). In contrast, auditory looming cues showed a different pattern of RTs across CTOA levels. Compared to a CTOA of 0 ms, RTs were significantly faster at both CTOAs of 250 ms ($t(14) = 4.471, p_{bonf} < 0.001, d = 1.154$) and 500 ms

$(t(14) = 2.971, p_{bonf} = 0.018, d = 0.767)$. Interestingly, RTs did not differ significantly between CTOAs of 250 ms and 500 ms $(t(14) = 1.500, p_{bonf} = 0.434, d = 0.387)$.

Discrimination sensitivity (d'; Macmillan and Creelman, 1991) were submitted to the same repeated measures ANOVA to determine if there were speed-accuracy tradeoffs across the cued conditions. There were no significant main effects for *Auditory Cue* ($F(3,42) = 0.304, p = 0.823, \omega^2 = 0.000$) and *CTOA* ($F(2,28) = 1.023, p = 0.372, \omega^2 = 0.002$). There was also no significant interaction for *Auditory Cue* and *CTOA* ($F(6,84) = 0.739, p = 0.620, \omega^2 = 0.000$).

The root-mean-squared-error (RMSE) of manual tracking during auditory cue presentation was also evaluated to determine if the cues impaired central task performance. There was no significant main effects of *Auditory Cue* ($F(3,42) = 0.938, p = 0.431, \omega^2 = 0.000$), *CTOA* ($F(2,28) = 0.533, p = 0.593, \omega^2 = 0.000$), or their interaction ($F(6,84) = 1.393, p = 0.227, \omega^2 = 0.025$).

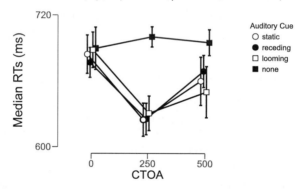

Figure 8: Interaction of *Auditory Cue* and *CTOA*. Cued reaction times were fastest for the CTOA level of 250 ms, relative to the simultaneous presentation (0 ms) of cue and target. This reaction time benefit decreases at 500 ms, particularly for static and receding cues. Error bars represent 95% confidence intervals.

The results of Experiment 1 reveal that auditory cues can induce a cross-modal reorienting of spatial attention that result in faster tilt-discriminations of peripheral targets. In a dual-task paradigm with a central task that demands attention continuously, this manifests itself as a response time benefit, with no influence on discrimination sensitivity and at no noticeable cost to central task performance. Generally, this crossmodal benefit is transient. It peaks when the visual targets appear shortly after auditory cue onset at a CTOA of 250 ms and diminishes with extended cue presentation at CTOA 500 ms. Contrary to our expectations, auditory looming cues did not exhibit cuing benefits that were stronger to either the static or receding cue. However, the cuing benefits of auditory looming cues do not diminish as quickly as static and receding cues. There continued to be a cuing benefit for auditory looming cues at a CTOA of 500 ms that did not significantly differ from the benefit observed at a CTOA of 250 ms. This persistent benefit could be due to the larger intensity of the auditory looming cue when the visual target appears. To investigate this, we performed a follow-up experiment that controlled for intensity levels.

4.4 EXPERIMENT 2: CAN THE SUSTAINED PERFORMANCE BENEFIT OF AN AUDITORY LOOMING SOUND AT LATE CTOAS BE ATTRIBUTED TO ITS HIGH INTENSITY WHEN THE VISUAL TARGET APPEARS?

In Experiment 1, auditory looming cues demonstrated a cuing benefit that was apparent even at the end of their presentation. In contrast, the cuing benefit of auditory static and receding cues is significantly diminished at the same time point. One reason for this difference could have been the higher intensity level of the auditory looming cue when the visual target appeared. To control for intensity levels, we introduced two additional auditory cues in Experiment 2. A *loud-static* cue that had the same intensity through its presentation as the offset intensity of the original looming cue and a *soft-looming* cue that had the same offset intensity of the original static cue. The original sounds were labeled *loud-looming* and *soft-static* in Experiment 2.

4.4.1 Results and Discussion

To begin, we performed one-tailed paired samples t-tests ($\alpha = 0.05$) and confirmed that all auditory cues induced a cuing benefit on visual targets (range of means=664–688 ms), compared to instances when targets were not preceded by an auditory cue (mean=736 ms; SE=79). Figure 9 summarizes the RTs across the conditions that presented an auditory cue. Auditory looming cues appear to induce a cuing benefit that does not change with intensity levels, while a loud-static cue induces a larger cuing benefit than a soft-static cue.

The median RTs of Experiment 2 (Figure 9) were submitted to a repeated-measures ANOVA for the factors of *Auditory Cue* (static, looming) and *Intensity* (soft, loud). There were no significant main effects of *Auditory Cue* ($F(1, 14) = 0.230, p = 0.639, \omega^2 = 0.000$) and *Intensity* ($F(1, 14) = 1.486, p = 0.243, \omega^2 =$

0.029). More importantly, this analysis revealed a significant interaction ($F(1, 14) =$ 7.305, $p = 0.017, \omega^2 = 0.283$), confirming our interpretation of the cuing benefits of the different auditory cues.

The soft-static cues and loud-looming cues were respectively equivalent to the static and looming cues employed in Experiment 1. With two-tailed paired samples t-tests, we found a significant difference between the cuing benefits of soft-static and soft-looming ($t(14) = 2.220, p = 0.043, d = 0.573$), and no significant difference between loud-looming and loud-static cues ($t(14) = 1.156, p = 0.267, d = 0.298$). From this, we inferred that auditory looming cues with a lower intensity would have continued to generate the sustained cuing benefit across different CTOAs, as observed in Experiment 1. In contrast, the cuing benefit of static cues at a CTOA of 500 ms can only be increased by raising their intensity levels.

The same analyses were performed on d' scores. Paired samples t-tests confirmed that none of the auditory cues improved discrimination sensitivity (range of means=1.680–1.781) relative to uncued trials (mean=1.653, SE=0.206). The ANOVA revealed no significant main effects and interactions for the factors of *Auditory Cue* and *Intensity*.

Similarly, we analyzed the RMSE on the visuo-motor tracking task to check for interference from auditory cues. Paired samples t-tests confirmed that none of the auditory cues significantly reduced central task performance (range of means=1.374–1.438), relative to comparable periods when no cue was presented (mean=1.407, SE=0.082). The ANOVA revealed no significant main effects and interactions for the factors of *Auditory Cue* and *Intensity*.

To summarize, we found that auditory looming cues were equally effective regardless of their intensity levels at a late CTOA of 500 ms. Thus, the pattern of cuing benefit observed in Experiment 1 is unlikely to have resulted from the cue's intensity when the visual target appeared.

4.5 DISCUSSION

Looming sounds are believed to be intrinsically salient (Neuhoff, 1998; Bach et al., 2009; Neuhoff, 2001). From this, we might expect them to operate optimally as transient cues, which involuntary capture attention to themselves and improve the discrimination of spatially congruent targets. Specifically, we expected a reflexive pattern of cuing benefit that is transient and diminishes with time (Müller and Rabbitt, 1989; Hein et al., 2006; Cheal et al., 1991; Nakayama

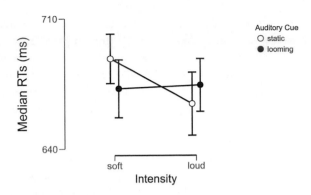

Figure 9: Interaction of *Auditory Cue* and *Intensity*. Median RTs of looming cues do not vary with *Intensity* levels. In contrast, loud static cues induce faster RTs than soft static cues. Error bars represent 95% confidence intervals.

and Mackeben, 1989). This pattern is verified with all auditory cues, regardless of sound type. We observed a general crossmodal benefit of all auditory cues, which hasten the accurate discrimination of a spatially congruent target that subsequently follows (i.e., 250 ms). However, we also found that auditory looming cues exert a stronger crossmodal influence on visual attention reorienting than static or receding cues, which only manifested at a later CTOA (i.e., 500 ms). We found that an auditory looming cue distinguishes itself by sustaining a large cuing benefit throughout its duration. In contrast, the crossmodal benefit of auditory receding and static cues is substantially diminished towards the end of their presentation (i.e., 500 ms CTOA). The follow-up experiment confirmed that this characteristic of an auditory looming cue's influence could not be directly attributed to the looming cue's intensity when the visual target appears; a soft looming cue sustained a cuing benefit for its entire duration as well as a loud looming cue.

This study was motivated to examine visual attention reorienting throughout the time course of an auditory cue. To do so, we investigated performance benefits at different time-points of the auditory cue's presentation as well as ensured that participants were continuously occupied with a central visuo-motor tracking task. In this regard, our experiment design differs from previous work that have similarly addressed the crossmodal influence of looming sounds on visual processing. To begin, previous studies have typically presented a visual target simultaneously with the onset of a looming sound (Leo et al., 2011; Cappe et al., 2009, 2012; Maier et al., 2004; Tyll et al., 2013) or after a looming sound has been presented (Bach et al., 2009, 2008). The former paradigm typically addressed supramodal influences of looming sounds on multisensory integration and the latter, aspects related to phasic alerting. Few studies (Leo et al., 2011) have directly investigated how looming sounds could serve as crossmodal cues in reorienting visual spatial attention. The current study is the first to outline the time course of this influence. Looming sounds exert a sustained reorienting benefit throughout their presentation while the crossmodal influence of comparable auditory cues wane with time.

At first glance, our results might appear to differ from those reported by Leo and colleagues (Leo et al., 2011). To recap, they reported that monoaural auditory looming cues induce preferential bias in tilt-discrimination sensitivity of the visual hemifield that is spatially congruent to the presentation ear, relative to the hemifield that is incongruent. We only found response time benefits when the auditory cues preceded visual targets and did not find a crossmodal benefit when auditory cues appeared simultaneously with the visual targets (i.e., 0 ms CTOA). Several differences exist between the current experiments and theirs. Critically, we sought to investigate how a crossmodal influence of auditory cues on spatially congruent visual targets might operate throughout their presentation. Therefore, we varied the CTOAs of auditory cues and peripheral visual targets, without varying spatial congruency. In contrast, Leo et al. (2011) evaluated the crossmodal influence of non-predictive sounds on simultaneously presented visual targets, which were either spatially congruent or incongruent. It should be noted that Leo et al. (2011) reported a significant interaction of sound type and spatial congruency, but not a main effect of either factors. In other words, their results were similar to our findings at 0 ms CTOA, namely that a looming sound or a spatially congruent sound does not in itself facilitate responses to a simultaneously presented visual target in the periphery—that is, not unless it is contrasted to performance in the opponent hemifield. Therefore, our findings are consistent with those of Leo and colleagues (Leo et al., 2011).

It was unexpected that we did not observe larger cuing benefits for looming sounds at the early CTOA levels of 0 and 250 ms. This was predicted given the known saliency of looming sounds. As mentioned previously, looming sounds induce larger skin conductance responses (Bach et al., 2009), are responded to faster (Cappe et al., 2009; McCarthy and Olsen, 2017; Skarratt et al., 2009), preferentially activate the amygdala (Bach et al., 2008) and the auditory cortex (Maier and Ghazanfar, 2007), and are subjectively rated as being more arousing (Bach et al., 2009). This could result in a crossmodal facilitation of visual processing, especially to stimuli that are spatially congruent and simultaneously presented, as has been noted in research on multisensory integration (Leo et al., 2011; Cappe et al., 2009, 2012; Maier et al., 2004; Tyll et al., 2013). One reason for this could be that our visual targets were tuned to a high discrimination threshold (i.e., 70%). Nonetheless, Leo and colleagues employed comparably tuned stimuli (i.e., 75%) and even reported that discrimination sensitivity were further increased during testing by the concurrent presentation of a spatially congruent looming sound. Thus, we do not believe that our results reflect ceiling effects. Leo and colleagues (Leo et al., 2011) explicitly emphasized response accuracy over speed (p. 196). Therefore, the performance benefits that they reported could have been influenced by looming sounds at any time-point after the onset of the accompanying sound, early or late. In contrast, our experiment required participants to continuously perform a central visuo-motor tracking task that introduced a time cost to attending to a peripheral location. This implicitly motivated participants to reorient their attention as quickly as possible but discouraged sustaining attention away from the central task for too long.

We believe that the cuing benefits of auditory looming cues reflect their sustained influence on reorienting spatial attention. Auditory looming cues oblige the visual system to continue paying attention to a region of the visual periphery that would otherwise remain neglected, especially when attentional resources are demanded elsewhere. As long as a sound increases in intensity, it is suggestive of an approaching object that will soon appear even if it is not yet apparent. While intensity is a salient property of sensory stimuli, its influence on attention is transient if it stays unchanged over time. Electrophysiological research in non-human primates have shown that looming sounds not only affect the primary auditory cortex, but also those areas that are involved in space recognition, auditory motion perception, as well as attention (Lu et al., 2001; Maier and Ghazanfar, 2007). Similarly, fMRI studies have also implicated a network of regions that are selective to looming sounds that are involved in the evaluation of complex object motion (i.e., superior and middle temporal sulcus; Seifritz et al., 2002; Bach et al., 2008). More recently, an MEG study have shown sustained neural activity in the right temporo-occipito-parietal junction and bilateral inferior temporal gyrus, which tracked the increasing intensity of looming sounds with long durations (i.e., 1600 ms). This resulted in significant differences against the neural activity generated by falling intensities of receding sounds, particularly at late periods (i.e., 900–1400 ms) after sound onsets. More interesting, this regions are not considered to be part of the auditory cortex, hence suggesting a supramodal influence of looming sounds on attention (Bach et al., 2015). Sustained activity in these regions suggest that looming sounds exert a continuous bottom-up influence on attention, as long as they continue to increase in intensity and indicate the potential approach of a relevant object.

In conclusion, the current study reports that auditory looming cues exert a crossmodal benefit on visual spatial attention that is sustained for its presentation duration. This is achieved even though spatial attention is constantly in demand by a central manual tracking task. The current results are believed to reflect a sustained vigilance during the looming sound's presentation and are potentially independent of its multisensory facilitation of visual processing per se. This benefit is founded on the rising intensity of the auditory looming cue and is independent of low-level characteristics (i.e., intensity) that can also influence an involuntary but transient influence on spatial attention. After all, the gentle prowl of a tiger can be as deadly as the clumsy stampede of cattle, but only when they near us.

4.6 METHODS

4.6.1 *Participants*

Thirty healthy volunteers participated in the current study (*Experiment 1*: 7 males, 8 females; mean age=26.67 years $\pm$ 4.78 s.d.; *Experiment 2*: 5 males, 10 females; mean age=24.67 years $\pm$ 3.79 s.d.). All participants reported normal hearing, normal (or corrected-to-normal) vision, and no history of neurological problems. They received written instructions, gave informed signed consent, and were remunerated 8 Euros/hour for their voluntary participation. The experimental procedure was approved by the Ethics Council at the University Hospital Tuebingen and carried out in accordance with their specified guidelines and regulations (see DOI 10.17605/OSF.IO/4WYGJ).

4.6.2 *Design*

Experiments 1 and 2 employed a full factorial repeated measures design. Experiment 1 had two independent variables with three levels each: (1) *CTOA* between an auditory cue and the peripheral visual target (either 0, 250, or 500 ms), (2) the intensity profile of the *Auditory Cue* (static, looming, and receding). Experiment

2 had two independent variables with two levels each: (1) intensity profile of the *Auditory Cue* (static, looming), (2) *Intensity* of the auditory cue (low, high). The *CTOA* in Experiment 2 was fixed at 500 ms. The primary dependent variable was the mean of the reciprocal response time for correct responses to the peripheral target.

Every session consisted of several 4.5 mins blocks of continuous manual tracking with 60 trials of a single-stimulus forced-choice (1AFC) peripheral tilt-discrimination task. Experiment 1 consisted of three sessions (15 blocks each) performed over consecutive days. *CTOA* was fixed for each block and the presentation order of *CTOA* was counterbalanced within sessions and across participants. Experiment 2 was conducted in one session (i.e., 20 blocks) on a single day.

4.6.3 Stimuli

For Experiment 1, the auditory cues were 400 Hz tones with triangular waveforms, created using the MATLAB sawtooth function sampled at 44.1 kHz. Their duration was 500 ms and their intensity over time were shaped to assume one of three profiles: looming, receding, or static. The looming sound was characterized by a dynamic increase from 30 to 75 dB SPL, as measured at the participants approximate head position. This change in intensity could be described as an audio object that approaches at the speed of 29 m/s towards the observer, with a time-to-contact of 500 ms (Gray, 2011). The receding sound was created by reversing the looming sound in time. The static sound had a non-changing intensity that was the mean intensity level of the looming (and receding) sound, i.e. 65 dB. To avoid clicking noise at sound on- and offsets, all sounds were convolved with a trapezoid grating such that they had 5 ms ramps at the sound on- and offset.

For Experiment 2, the static and looming sounds were modified to create versions that ended with comparable low and high intensities. In Experiment 2, the original looming sound was regarded as the loud-looming cue and the original static sound as the soft-static cue. Accordingly, loud-static cue was the static sound with an adjusted intensity that matched the loud-looming cue's end-intensity of 75 dB, while a soft-looming cue was a looming sound that ended with an intensity of the original static sound 65 dB. These intensity profiles are visualized in Figure 10B. All auditory stimuli can be accessed at DOI 10.17605/OSF.IO/4WYGJ.

In both experiments, the peripheral visual targets were Gabor patches of 2 degrees visual angle in diameter, with a spatial frequency of 3.1 cycles/degree and a contrast of 50% (background gray = 20.3 cd/m^2, Gabor patch black = 1.3 cd/m^2, Gabor patch white 38.3 cd/m^2). They were always presented 9° to the left or right of the manually tracked crosshair for 250 ms. A pre-testing adaptive procedure tuned the orientation tilts to be 70% orientation discrimination threshold of each participant (mean threshold and standard error for the left hemifield: 1.74° $\pm$ 0.27, and the right hemifield: 1.70° $\pm$ 0.21; Levitt, 1971).

A compensatory visuo-motor tracking task, presented in the center, had to be performed continuously. A crosshair cursor comprising a vertical and horizontal line (0.70° long) was continuously and vertically displaced from a dotted horizontal reference line (5.43° long), along the vertical screen center. Participants rejected this cursor displacement to stabilize this cursor by deflecting a joystick forwards and backwards, which controlled the cursors vertical velocity and acceleration with equal weighting. In the absence of manual inputs, the cursor displacement was controlled by a quasi-random reference signal that was a sinusoidal function comprised of a sum of ten, non-harmonically related sine waves. This function had a variance of 1.62° (Nieuwenhuizen et al., 2013).

4.6.4 *Apparatus*

The experiment was controlled with custom-written software in MATLAB 8.2.0.701 (R2013b) and Psychophysics Toolbox 3.0.12 (Brainard, 1997; Pelli, 1997; Kleiner et al., 2007). A ViewPixx Screen (60.5 x 36.3 resolution; 120 *Hz*) presented all visual stimuli, at a fixed distance of 45 cm from chin-rest. Sound presentation was controlled by an ASIO compatible sound card (SoundBlaster ZxR; Creative Labs) and presented monophically through either the left or right speaker of a pair of headphones (MDR-CD380; Sony). The right and left arrow inputs of a standardized keyboard were used for collecting left and right tilt discrimination responses respectively. A right-handed control stick (Hotas Warthog Flight Stick) was used for the central manual tracking task.

4.6.5 *Procedure*

Prior to testing and after experiment briefing, participants performed five practice blocks of manual tracking only, followed by an adaptive procedure on a 1AFC tilt-discrimination task on peripheral visual targets (Levitt, 1971). The adaptive procedure determined the tilt that corresponded to the participant's 70% orientation discrimination threshold. Participants fixated a static central cross through this adaptive procedure. To determine individual thresholds, we employed a 1-up-2-down staircase procedure with six interleaved staircases, evenly divided for the left and right hemifields. The vertical tilt of the Gabor stimuli had starting values of 0.0°, 2.5°, and 5.0° for three staircases per hemifield. Each staircase allowed for a maximum of 100 trials or terminated after 19 reversals, whichever came first. The first four reversals had the respective step sizes of 1.0°, 0.5°, 0.25°, and 0.1°, which then remained constant for the rest of the adaptive procedure. This was always performed in the first experimental session.

Upon completion, participants were allowed to perform the test blocks (Experiment 1: n=45; Experiment 2: n=20). Participants were required to perform two concurrent tasks on every test block: (1) a compensatory manual tracking task on a central crosshair, (2) a 1AFC tilt-discrimination task on peripheral visual targets. Mandatory 1.5 min rest breaks were provided between blocks.

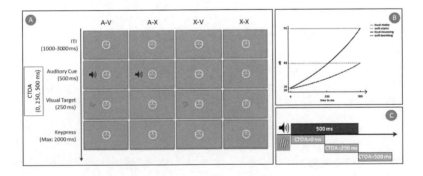

Figure 10: Experiment procedure and stimuli. (A) Four instances of trials of equal probability that could require participants to perform tilt-discrimination on a peripheral visual target (not drawn to scale for visibility). (B) Auditory cues used in Experiment 2, whereby the soft-static cue and loud-looming cue were the static and looming cue of Experiment 1. (C) Visual targets could appear at the onset of the auditory cue or after the onset.

In the compensatory tracking task, participants deflected the right-handed joystick in either the forwards or backwards direction to their body in order to counteract movements of the crosshair in either the upwards or downward direction respectively. The goal was to stabilize a central crosshair on a horizontal dotted line. In the tilt-discrimination task, participants had to determine the tilt of peripheral targets when they appeared on either the left or right side of the crosshair. They responded by using their left index or ring finger to respectively indicate a left or right diagonal tilt. Participants were instructed to maintain fixation of the central crosshair of the manual tracking task throughout the experiment.

Trials occurred every 2000 ms ± 1000 ms (uniform distribution) and presented either an auditory cue only (A-X), a peripheral visual target only (X-V), an auditory cue followed by a peripheral visual target (A-V), or neither cue nor target (X-X). This ensured that the auditory cue was non-predictive of target appearance. When an auditory cue was presented, they were always presented via the headphone (i.e., right/left) that was on the same side as the upcoming visual target. When visual targets were presented, they appeared equally often on the left and the right of the central visuo-motor tracking task. In Experiment 1, they could occur at the onset of an auditory cue (if any), or 250 ms or 500 ms after the cue onset (see Figure 10C). In Experiment 2, visual targets always appeared 500 ms after the auditory onset. A fixed duration of 2000 ms for keypress responses always took place after a visual target was supposed to be presented. The timelines of these four possible trials are illustrated in Figure 10A.

After completing the required number of test blocks, participants were debriefed on the purpose of the experiment.

NEURAL CORRELATES FOR THE AUDITORY CUE BENEFIT

This chapter has been reproduced from an article that is in preparation for publication: Glatz, C., Miyakoshi M., Bülthoff, H. H., and Chuang, L. L. (2018). rP280: An ERP marker for the looming cue benefit and its underlying neural networks.

5.1 ABSTRACT

Previous work has demonstrated preferential processing of looming sounds in terms of behavior as well as neural activity. While these results favor looming sounds as auditory cues, surprisingly little work on the utility of auditory looming cues in shifting visuo-spatial attention in dual-tasking has been done.

The current study used auditory looming and static cues to shift attention from a central ongoing task to an occasional peripheral target. We found that auditory looming cues bring about faster correct discrimination of peripheral visual targets than static cues. Additionally, we identified an electrophysiological marker for this looming cue benefit in shifting visuo-spatial attention, the rP280. Through source estimation analysis we identified two neural networks that gave rise to the looming cue benefit. The posterior network responded more to looming than static cues which was further amplified by reduced inhibition for looming cues in the coexisting frontal network.

In summary, the current results suggest that the looming saliency extends from preferential processing to auditory looming cue benefits. Looming sounds that precede a target might be better cues than, for example, static sounds for shifting attention away from an ongoing attention-capturing task.

5.2 INTRODUCTION

In the animal kingdom, it is often vital to one's survival to be distracted. The appearance of certain sensory information can readily indicate the approach of a predator (or prey) that demand a time-critical disengagement from current activity (Tinbergen, 1951). Some sensory events are more salient than others. In particular, it is well-established that looming stimuli, namely visual or auditory events that increase in size or intensity, benefit from preferential processing in a way that is evident in both behavior (e.g., Neuhoff, 1998; Leo et al., 2011; Schouten et al., 2011; Sutherland et al., 2014; Cecere et al., 2014) as well as brain activity (e.g., Maier and Ghazanfar, 2007; Romei et al., 2009). Previously, we reported that looming sounds distinguish themselves from other sounds, not by redirecting attention away from an ongoing central activity but by sustaining this redirected attention for the duration of its presentation (Glatz and Chuang, 2018, under review). In contrast, other comparable sounds redirected attention that diminished rapidly even while they were still present. In this paper, we investigate the neural mechanisms that could underlie this cuing benefit of auditory looming cues.

Looming sounds are perceived as especially salient because the rising intensity over time signals an approaching object (Middlebrooks and Green, 1991; Guski, 1992) that could pose a threat. For example, looming sounds are perceived as moving faster (i.e., change more in intensity) than receding sounds (departing object; Neuhoff, 1998, 2001; Olsen and Stevens, 2010; Seifritz et al., 2002; Neuhoff, 2016). This bias in perception results in the consistent underestimation of the time of arrival for looming sound sources (Neuhoff et al., 2009, 2012; Riskind et al., 2014; Rosenblum et al., 1987, 1993; Schiff and Oldak, 1990; Ashmead et al., 1995). That is, listeners perceive the looming sound source closer than it actually is and, hence, anticipate an earlier than actual arrival. The advantage of this anticipatory behavior has a potentially large gain, giving more time than expected to prepare a confrontation or avoidance of the approaching object. Erring on the side of caution and providing more preparation time also provides a margin of safety. Interestingly, vulnerable individuals show a stronger looming bias, plausibly because they consider themselves in need of a larger margin of safety (Neuhoff et al., 2012; Riskind et al., 2014; McGuire et al., 2016).

The simultaneous presentation of looming sounds with visual stimuli has been shown to influence the visual processing (Leo et al., 2011; Schouten et al., 2011; Sutherland et al., 2014). On a behavioral level, looming sounds were found to bias visual stimuli such that static visual stimuli were judged as brighter and larger while depth-ambiguous visual stimuli were perceived as facing the viewer (Sutherland et al., 2014; Schouten et al., 2011). This could be because the approaching motion conveyed by the looming sound might be remapped onto visual coordinates and, therefore, alter the visual perception. On a neural level, looming sounds were found to enhanced visual cortex excitability (Romei et al., 2009) but also preferentially activate neural regions for multisensory integration of the looming property (Cappe et al., 2012).

In contrast, surprisingly little work has investigated looming sounds as cues to indicate an approaching (visual) target. The beneficial effect for presenting visuals together with looming sounds cannot be attributed to the looming property alone but might be confounded by a multisensory benefit. Studies that presented a looming sound to cue a visual stimulus have consistently found faster response times to the visual stimulus when cued by an approaching compared to receding or static sound (Bach et al., 2009; Burton, 2011; Ho et al., 2013; Glatz and Chuang, 2018, under review). The faster response times to looming cued visual targets could be due to increased phasic arousal. However, previous work has shown that looming sounds enhance attention that is location specific (Leo et al., 2011). This might be due to the fact that looming sounds communicate the object's approaching trajectory (Middlebrooks and Green, 1991; Guski, 1992). For this reason, looming sounds might be especially well suited in directing visuo-spatial attention to neglected locations in space. Directing attention in advance to the target's location might result in enhanced neural processing and, thus, improved performance.

All this evidence favors the looming sound as an auditory cue for directing visuo-spatial attention. Previous work that supports this finding has focused on single task effects. Although, these might not generalize to dual-task paradigms. Santangelo and Spence (2007) demonstrated that increasing the demands of a primary task could eliminate spatial cuing effects in a secondary task. This suggests that the auditory looming cue benefit for visual targets cannot readily be transfered to dual-tasking, since it has mainly been investigated in isolation without any distracting tasks. Research on visual looming in dual-tasking has shown that looming objects capture and direct attention away from an ongoing task to the visual looming objects (King et al., 1992; Franconeri and Simons, 2003; Terry et al., 2008). If we assume that looming objects capture attention automatically in a stimulus-driven way, the responses to such approaching objects should not be affected by cognitive load (Engström et al., 2017). Indeed, looming sounds have been used effectively to indicate an impending frontal collision while driving. Gray (2011)'s findings showed that participants braked significantly earlier when presented a looming sound compared to comparable auditory warnings. Another instance that supports the looming sound's effectiveness in dual-task scenarios tested whether the auditory looming bias in estimating arrival time was affected by the cognitive load imposed by another tasks (McGuire et al., 2016). Their results showed an even larger looming bias under high load, underestimating the time of arrival even more, than under low cognitive load. This suggests that auditory looming cues can efficiently disengage listeners from their current task and direct their attention to the looming object without much effort. Despite these encouraging results, suggesting an auditory looming benefit in dual-task, neither of the two experiments required a shift of spatial attention.

In the present study, we will demonstrate that a looming sound is an effective crossmodal cue for reorienting visuo-spatial attention from a central ongoing task to an occasional peripheral target. Furthermore, we identify the neural network that underlies this looming benefit. Our experimental paradigm was explicitly designed to address the two aforementioned research gaps. To begin, we ensured that participants' attention was always engaged in a central visuo-motor tracking task, as opposed to mere central fixation. Peripheral visual targets were occasionally presented for identification that could be preceded, not accompanied, by an auditory sound of static or looming intensity. Thus, any benefit of looming sounds was due to crossmodal attentional cuing and not multisensory salience. We recorded participants' response times to identify visual targets and expected faster responses when cued by a looming sound. Control analyses were performed to ensure that facilitation in response times were not accompanied with eye-movements. We also analyzed the ERP response to visual stimuli and, from this, identified the neural regions that could underlie this crossmodal looming cue benefit.

5.3 METHODS

This study compared looming and static auditory cues with different intensities for their effectiveness in orienting attention to peripheral visual stimuli. The experiment was a within-subject repeated measures design with the independent variables of *Sound Type* (static, looming) and *Intensity* (soft, loud). The dependent variables used to evaluate the auditory cues were response times, discrimination sensitivity, and ERP components.

5.3.1 Participants

Twenty healthy volunteers (mean age = 25.10 ± 3.81; 7 males) were recruited from a subject data base. All participants reported normal or (corrected-to-normal) vision, no known hearing deficits, and no known history of neurological problems. All participants gave signed consent to written instructions and were remunerated 12 Euros/hour for their voluntary participation. The experimental procedure was approved by the ethics council of the University Hospital Tuebingen.

5.3.2 Stimuli

Central task

To ensure that participants' attention was always engaged in the center, they had to continuously perform a visuo-motor control task. For this, participants counteracted the cursor displacement by deflecting the joystick forward and backward to stabilize the cursor in the display's center. The cursor for this compensatory manual tracking task was $0.7°$ long and moved vertically about the display's center which was depicted by a dotted horizontal reference line ($5.43°$ long). The manual manipulation of the joystick controlled the cursor's vertical velocity and acceleration with equal weighting. Without manual input, the cursor's vertical position was determined by a quasi-random reference signal, which was comprised by a sum of ten non-harmonically related sinusoids. The disturbance function had a variance of $1.62°$ and was adapted from Nieuwenhuizen et al. (2013). The amplitude $A(j)$, frequency $\omega(j)$, and phase $\phi(j)$ of these 10 sinusoids are provided in Table 3. This task was presented on every block and lasted 4.5 min.

j	1	2	3	4	5	6	7	8	9	10
$A(j)$ in deg	1.34	1.03	0.51	0.26	0.16	0.10	0.06	0.04	0.04	0.03
$\omega(j)$ in rad/s	0.39	0.83	1.75	2.83	3.94	5.43	7.73	10.52	13.12	17.33
$\phi(j)$ in rad	0.15	2.07	1.02	5.58	0.99	0.63	3.45	5.40	5.44	3.38

Table 3: Amplitude, frequency, and phase for 10 sinusoids constituting the reference signal

Peripheral task

The auditory cues (duration: 500 ms) used were 400 Hz triangular waveforms, sampled at 44.1 kHz. These sounds were presented with either a static or looming intensity profile over time, at two levels of intensity (see Table 4). The loud looming sound had a exponentially rising intensity profile (30-75 dB), signaling the time-to-contact, according to Gray (2011). Two static intensity sounds were created to control for (1) the average intensity of the loud looming sound (static soft, 65 dB) and (2) the end intensity of the loud looming sound (static loud, 75 dB). The soft looming sound (20-65 dB) had the same intensity profile as the loud looming sound but ended in the intensity of the soft static sound. The sound pressure levels were measured with an audiometer at participants' approximate head position. To avoid clicking noise at sound on- and offsets, all sounds were multiplied by a trapezoid grating such that they had 5 ms ramps at the on- and offset.

	static	looming
soft	65 - 65	20 - 65
loud	75 - 75	30 - 75

Table 4: Intensity in dB for the static and looming sounds with two intensity levels.

The visual stimuli were tilted Gabor patches of 2 degree visual angle in diameter, with a spatial frequency of 3.1 cycles/degree and a contrast of 50% (background gray = 20.3cd/m^2, Gabor patch black = 1.3cd/m^2, Gabor patch white 38.3cd/m^2). They were presented at an eccentricity of 9 degrees visual angle for 250 ms. The tilt of the Gabor patch was tuned individually to 70% orientation discrimination threshold of each participant (see Section 5.3.4).

5.3.3 Apparatus

The experiment was conducted in a dark room which was insulated against external sounds. A desktop display (ViewPixx Screen, 60.5 x 36.3 resolution; 120 Hz) was used to present the visualization at a fixed distance of 45 cm from the participant, who was in a chin-rest. Customized software (MATLAB 8.2.0.701, R2013b) and Psychophysics Toolbox 3.0.12 (Brainard, 1997; Pelli, 1997; Kleiner et al., 2007) controlled the experiment and data collection. Sound presentation was controlled by an ASIO 2.0 compatible sound card (SoundBlaster ZxR; Creative Labs). The auditory stimuli were presented via stereo headphones (MDR-CD380; Sony). Participants used a USB control side stick (Hotas Warthog Flight Stick; Thrustmaster) for visuo-motor tracking task. Participants used the left and right arrow keys of a standardized USB keyboard to indicate the Gabor patch's tilt.

5.3.4 Procedure

The experiment consisted of two sessions, conducted on separate days. The first session included two additional parts. First, participants familiarized themselves with the visuo-motor tracking task in 5 practice trials prior to data collection. Subsequently, participants' individual 70% correct orientation discrimination threshold was determined by an adaptive 1-up-2-down staircase procedure (see Glatz and Chuang, 2018, under review, for more details). The result of this visual-orientation threshold procedure determined the tilt of the Gabor patches for every participant individually (mean threshold left hemifield: $1.62° \pm 0.82$, right hemifield: $1.73° \pm 0.76$).

In both sessions, participants performed the same experiment that involved a continuous tracking task and an intermittent tilt discrimination task. To bring the cursor to the designated center of the screen, participants had to continuously move a joystick forward and backward with their right hand. In the tilt discrimination task, participants had to indicate the tilt direction of peripheral visual targets (i.e. Gabor patches) as fast and accurately as possible, whenever they appeared. To respond, they had to press the left arrow (left ring finger) for the anti-clockwise (left) tilt and the right arrow (left index finger) for the clockwise (right) tilt, without moving their eyes from the cursor.

Each session consisted of 15 experimental blocks (4.5 min per block) separated by mandatory 1.5 min breaks. Each block, contained 60 trials of the discrimination. One of four equally probable trial types could be presented (see Figure 11): (1) trials that presented an auditory cue were followed by a visual stimulus (AV), (2) trials presented an auditory cue but no visual stimulus (AX), (3) trials presented a visual stimulus without a preceding auditory cue (XV), (4) trials that presented neither auditory cue nor visual stimulus (XX). Independent of the type, each trial was followed by a 2000 ms response interval and trials were separated by an inter-trial-interval (ITI), randomly sampled from a uniform time range of 1000-3000 ms. It is worth mentioning that having these four conditions, the auditory cues were never predictive of a visual stimulus appearing. Nonetheless, in AV conditions the auditory cue (A) and the visual stimulus (V) were always presented from the same side. Hence, A's location was always predictive of V's location.

Following aspects were controlled: An equal number of auditory cues was presented on AV and AX trials in random order. The visual stimulus' tilt and position were pseudo-randomized. That is, 50% of the visual stimuli were presented in the right periphery of which 50% had a right and the other 50% a left tilt. The same applied to the 50% presented in the left periphery.

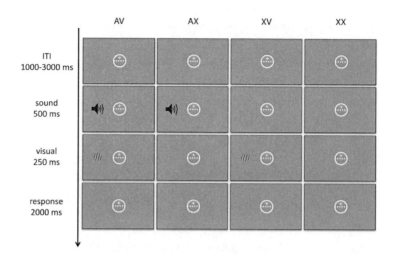

Figure 11: The trial sequence according to the condition. Trials were separated by an ITI of 1000-3000 ms which was followed by a 500 ms interval where an auditory cue could be presented. In the subsequent 250 ms a visual stimulus could be presented. Finally, participants had 2000 ms to respond to a presented visual stimulus. The auditory and visual stimulus were always presented from a congruent location. The presentation location, as well as the tilt direction, were counterbalanced such that a left sound preceded a left visual with left tilt as often as a left visual with a right tilt. Similarly, a right sound preceded a right visual with left tilt as often as a right visual with a right tilt.

5.3.5 EEG recording

Electroencephalography (EEG) recording were obtained from 59 active electrodes mounted to the scalp using an elastic cap according to the international 10-20 system (ActiCap System, Brain Products GmbH, Gilching, Germany). Horizontal and vertical electrooculogram recordings were obtained from four additional electrodes placed at the right and left canthi as well as above and below the left eye. All signals were recorded at a sampling rate of 1000 Hz, online referenced to FCz and grounded to AFz. By applying electrode gel to each electrode, we were able to ensure an impedance below 20 kOhm. Each participant's individual electrode positions were digitized using the CapTrak scanner (Brain Products GmbH, Gilching, Germany). Experimental events, such as the visual stimulus onset, were synchronized between EEG recording PC and experimental PC via a parallel connection.

5.3.6 EEG signal processing

Processing of the EEG signal was performed with EEGLAB v.14.0.0 (Delorme and Makeig, 2004) and ERPLAB v.6.1.3 (Lopez-Calderon and Luck, 2014) using MATLAB (8.2.0.701, R2013b). First, every data set was preprocessed according to the following steps. The data was downsampled to 250 Hz to reduce computational costs. Then, a high-pass filter (cut-off = 0.5 Hz) was applied to the data to remove any slow drifts. 50 Hz electrical noise from the environment contaminating the electrical brain activity was removed using CleanLine, a plugin in EEGLAB. Using artifact subspace reconstruction, bad channels, such as channels with flat lines, were removed. After cleaning the data, offline re-referencing to the common average reference was performed. Subsequently, the data was submitted to the Adaptive Mixture Independent Component Analysis (AMICA, Delorme et al., 2012) which decomposes the electrical activity recorded at electrodes into activity of independent components (ICs). Using a MNI Boundary Element Method head model, ICs were fit to equivalent dipoles (Piazza et al., 2016). IC dipoles with a residual variance larger than 15% and dipoles located outside the brain were excluded. Based on the power spectrum, the remaining ICs of all participants were grouped into 30 clusters using k-means. These clusters were then inspected for non-cortical electrical activity (e.g. eye-related activity, muscle-related activity, line noise) and unresolved components. Non-cortical activity was identified based on the clusters' power spectrum, their dipole location, and their scalp topography. After removing this non-cortical activity from the EEG data (see Chuang et al., 2017, for examples), the remaining cortical activity was backprojected to the sensor level (electrodes). Finally, the data was epoched by extracting time-windows (-500:1000 ms) around events (e.g., visual stimulus onset and auditory cue offset), baselined to the 500 ms before the event and averaged across all epochs for each specified condition.

5.4.1 Behavioral performance

Participants' performance in the discrimination was assessed by response times to correctly identified visual stimuli (RT) and discrimination sensitivity for visual stimuli (d'). To calculate the discrimination sensitivity for visual stimuli, hits and false alarms for the different conditions were assessed (Macmillan and Creelman, 1991). Tracking performance was assessed by calculating the root-mean-squared-error (RMSE) between the cursor's deviation from the designated center across 250 ms during visual stimulus presentation.

The analysis of the behavioral data consists of two steps. First, we analyzed whether cuing a visual target influenced the participants' performance in one of the three measures. In a second step, we submitted the respective measurement data to a 2 x 2 x 2 repeated measurements analysis of variance (ANOVA) with the factors *Sound Type* (static, looming), *Intensity* (soft, loud), and *Session* (1, 2).

INFLUENCE OF CUING First, we inspected whether the auditory cues improved discrimination sensitivity and response times by comparing the trials where an auditory cue preceded the visual stimulus to those were it did not. We found no significant difference for d' across the two conditions (1-tailed, paired-sample t-test; $t(19) = -.73, p = .24, d = -.16$) or tracking performance (1-tailed, paired-sample t-test; $t(19) = .60, p = .28, d = .13$). However, participants responded significantly faster to the visual stimulus when cued by a sound (1-tailed, paired-sample t-test; $t(19) = 11.01, p < .001, d = 2.46$).

To summarize, participants responded significantly faster to the visual discrimination task when there was a cue compared to when there was no auditory cue.

PERFORMANCE ON CUED TRIALS We performed a separate 2 x 2 x 2 ANOVA for each of the three metrics (RT, d', RMSE).

For d', we found no main effect of *Sound Type* ($F(1,19) = .26, p = .62, \omega^2 = .00$) and *Session* ($F(1,19) = 1.74, p = .20, \omega^2 = .03$), but for *Intensity* ($F(1,19) = 16.17, p < .001, \omega^2 = .42$). Post-hoc analysis showed that discrimination sensitivity was improved for loud auditory cues ($t(19) = -4.02, p < .001$). There were no significant interactions [*Sound Type* x *Intensity* ($F(1,19) = .32, p = .58, \omega^2 = .00$); *Sound Type* x *Session* ($F(1,19) = .09, p = .77, \omega^2 = .00$); *Inten-

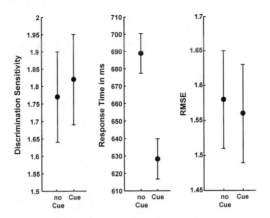

Figure 12: Behavioral performance for visual targets with and without a cue. The left shows the statistically not significant difference for discrimination sensitivity after not cued and cued visual targets. The middle shows the significantly faster correct response times for cued vs. not cued visual targets. The right shows the tracking performance during cued vs. not cued visual targets. Error bars display the 95% confidence intervals of the effect.

sity x Session ($F(1, 19) = 1.71, p = .21, \omega^2 = .03$); *Sound Type* x *Intensity* x *Session* ($F(1, 19) = .65, p = .43, \omega^2 = .00$)].

For RT, we found a main effect of *Sound Type* ($F(1, 19) = 6.61, p = .02, \omega^2 = .21$). Post-hoc analysis showed that looming cues resulted in faster response times than static cues ($t(19) = 2.57, p = .02$). There was no main effect of *Intensity* ($F(1, 19) = 2.16, p = .16, \omega^2 = .05$), but a main effect of *Session* ($F(1, 19) = 9.37, p = .01, \omega^2 = .29$). Post-hoc analysis indicated that participants were faster in the second session ($t(19) = 3.06, p = .01$). The only significant interaction was *Intensity* x *Session* ($F(1, 19) = 5.19, p = .04, \omega^2 = .17$). Post-hoc analysis showed that there was an effect of intensity in the first session ($t(19) = 2.38, p = .03, d = .53$) but not in the second session ($t(19) = -.77, p = .45, d = -.17$). No other interaction was significant [*Sound Type* x *Intensity* ($F(1, 19) = .34, p = .57, \omega^2 = .00$); *Sound Type* x *Session* ($F(1, 19) = .03, p = .86, \omega^2 = .00$); *Sound Type* x *Intensity* x *Session* ($F(1, 19) = .17, p = .68, \omega^2 = .00$)].

For RMSE, we found no main effect of *Sound Type* ($F(1, 19) = .11, p = .75, \omega^2 = .00$) or *Intensity* ($F(1, 19) = .13, p = .73, \omega^2 = .00$), but a main effect of *Session* ($F(1, 19) = 2.16, p < .001, \omega^2 = .47$). Post-hoc analysis showed that participants were better in the manual control task in the second session than in the first ($t(19) = 4.42, p < .001$). Only the three-way interaction (*Sound Type* x *Inten-*

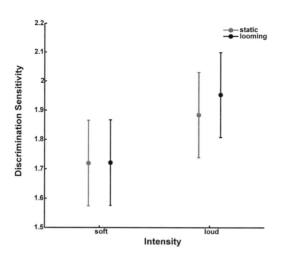

Figure 13: The influence of intensity on target discrimination sensitivity. Participants were identified visual targets significantly better that were cued by loud sounds. Error bars display the 95% confidence intervals of the main effect of intensity.

sity x Session) was significant ($F(1, 19) = 7.01, p = .02, \omega^2 = .22$). No other interaction was significant [*Sound Type x Intensity* ($F(1, 19) = .95, p = .34, \omega^2 = .00$); *Sound Type x Session* ($F(1, 19) = 1.00, p = .33, \omega^2 = .00$); *Intensity x Session* ($F(1, 19) = .84, p = .37, \omega^2 = .00$)].

To summarize, despite the fact that cues did not improve discrimination sensitivity, comparing different cues' effectiveness, d' improved only with increased sound intensity. Also tracking performance was not affected by sound type, however, it improved with time. Response times were significantly faster to visual stimuli cued by a looming sound independent of intensity. In addition, participants improved in the discrimination over time which interacted with intensity such that loud sounds sped up response times in the first session but not in the second session.

EYE MOVEMENTS   To ensure that participants only shifted covert and not over attention, we analyzed whether they moved their eyes to the visual target or the auditory cue. We found that there was a main effect of *Stimulus* (auditory cue, visual target) ($F(1, 19) = 9.57, p = .01, \omega^2 = .29$), a main effect of *Sound Type* (noSound, Constant Soft, Constant Loud, Looming Soft, Looming Loud)

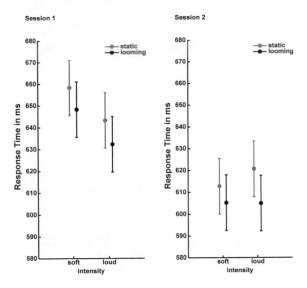

Figure 14: Correct response times to visual targets. The results demonstrate the main effect *Sound Type* and *Session*, as well as the interaction of *Intensity* x *Session*. Participants were significantly faster for looming sounds and improved with experience. Louder intensity initially also evoked faster response times but this effect diminished with time. Error bars display the 95% confidence intervals of the interaction effect.

$(F(4,76) = 4.98, p = .001, \omega^2 = .16)$, and a significant interaction *Sound Type* x *Stimulus* $(F(4,76) = 3.67, p = .01, \omega^2 = .12)$. This significant interaction can be explained by the fact that all visual targets preceded by an auditory cue generated fewer eye movements than those visual stimuli that were not cued by a sound, post hoc Tukey HSD comparisons confirm this (see Figure 15).

In summary, participants made more eye movements to the visual target when it was not cued compared to cued visual targets or auditory stimuli.

5.4.2 EEG/ERP responses

EEG ANALYSIS   We adopted a three stage approach to analyze the influence of auditory cues on the visual ERP. First, we examined the topography of the cued visual ERP, especially at posterior electrodes. Since we noticed a hemispheric difference, we identified the left and right parietal/occipital electrodes

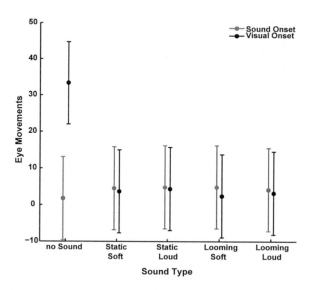

Figure 15: Mean eye movements according to condition. Participants made almost no eye movements to auditory cue onset or cued visual target onset. Only when visuals were not cued by a sound, participants made significantly more eye movements. The *no Sound* condition for sound onset reflects random eye movements while performing only the visuo-motor tracking task. Error bars display the 95% confidence intervals of the interaction effect of *Sound Type* x *Stimulus*.

as regions of interest. Second, we statistically analyzed the cue benefit for different auditory cues. That is, the processing that is due to having a cuing effect on a target without the processing of the cue or target. For the statistical analysis, a time window was determined by the peak in the grand average across all four auditory cue conditions, 280 ± 30 ms (based on the midline electrodes Pz, POz and Oz). Mean amplitudes for the different conditions and regions were submitted to a 2 x 2 x 2 ANOVA with the factors *Sound Type* (static, looming), *Intensity* (soft, loud), and *Hemisphere* (left, right). The third and last analysis step estimated the brain regions that gave rise to this ERP difference by performing a source localization analysis. For this purpose, the ICs of the stimulus-aligned (epoched) data were grouped into 15 clusters based on their dipole locations using k-means. Using the EEGLAB plugin *std_erpStudio*, we were able to determine the clusters that contributed to the observed difference and, therefore, estimate the cortical areas that gave rise to the effect. This was tested statistically by using a permutation test that provides the p-value, as well as, the number of unique subjects and ICs contributing to this cluster.

TOPOGRAPHY   Since all conditions contain equivalent numbers of epochs, our design allowed us to subtract ERP activity from the AX condition from ERP activity from the AV condition. By subtracting activity related to the processing of the auditory cue from activity generated by the combined presentation of auditory cue and visual stimulus, we can investigate the visual stimulus processing only. In other words, by means of this subtraction (AV-AX), we are able to analyze the effect auditory cues had on the processing of the visual stimulus, without mingling it with activity associated to sound processing. The ERPs displayed in Figure 16 are waveforms for cued visual stimuli (AV-AX; in black) and uncued visual stimuli (XV; in gray). When comparing these two conditions, the cued visual stimulus processing differs from the uncued visual stimulus processing in that its maximum peak appears earlier than for uncued visual stimulus processing (see Glatz et al., 2016). Also, it is noteworthy that there is a hemispheric difference at posterior electrodes for the cued visual stimulus processing which we, therefore, included into the statistical analysis.

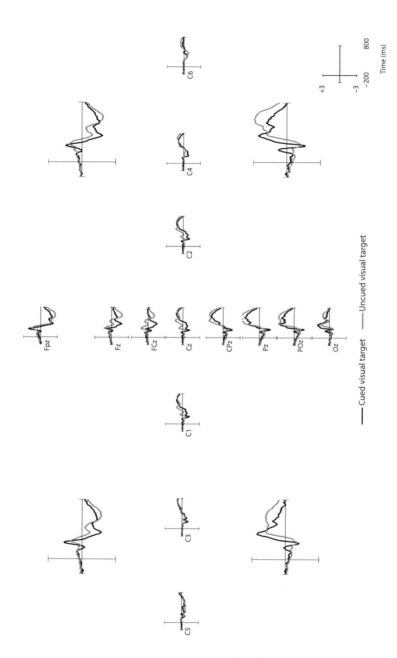

Figure 16: Visual target processing (target onset at 0 ms). The ERP waveform for cued (in black) and uncued (in gray) visual targets are displayed across the vertical and horizontal midline of electrodes, as well as averaged for anterior (Fp1, Fp2, AF3, AF4, AF7, AF8, F1, F2, F3, F4, F5, F6, F7, F8) and posterior (P1, P2, P3, P4, P5, P6, P7, P8, PO3, PO4, PO7, PO8, O1, O2) righ and left electrodes. Cued visual targets showed an earlier posterior maximum peak (Glatz et al., 2016) and a right lateralization.

ERP ANALYSIS OF VARIANCE   To investigate the benefit auditory cues had on the visual processing, we subtracted the activity related to auditory processing only (AX) and the activity related to uncued visual stimulus processing (XV) from the activity associated with the combined processing (AV-AX-XV). Figure 17 A and B show the ERP waveform for the auditory cue benefit for the four different sounds. The analysis of the ERP activity around 280 $\pm$30 ms (rP280) resulted in a main effect of *Sound Type* ($F(1,19) = 6.18, p = .02, \omega^2 = .20$). Post-hoc analysis showed that looming cues resulted in larger mean amplitudes than static cues ($t(19) = 2.48, p = .02$). There was no main effect of *Intensity* ($F(1,19) = .28, p = .60, \omega^2 = .00$) and no main effect of *Hemisphere* ($F(1,19) = .16, p = .69, \omega^2 = .00$). The only significant interaction was *Sound Type* x *Hemisphere* ($F(1,19) = 7.99, p = .01, \omega^2 = .25$). Post-hoc analysis showed that there was no significant difference of *Sound Type* in the left hemisphere ($t(19) = -1.39, p = .09, d = -.31$) but a significant difference of *Sound Type* in the right hemisphere ($t(19) = -3.25, p < .01, d = -.73$). No other interaction was significant [*Sound Type* x *Intensity* ($F(1,19) = .37, p = .55, \omega^2 = .00$); *Intensity* x *Hemisphere* ($F(1,19) = .53, p = .48, \omega^2 = .00$); *Sound Type* x *Intensity* x *Hemisphere* ($F(1,19) = .89, p = .36, \omega^2 = .00$)].

SOURCE LOCALIZATION   To see which source(s) might have given rise to the looming cue benefit, we estimated the dipoles that contributed to the looming rP280 benefit. Since ICs from different datasets cannot be combined, the analysis considered 2 x 20 subject datasets as 40 subjects. In the time interval 280 $\pm$ 30 ms, six dipole clusters showed a significant difference between looming and static cues (see Figure 18).

The three clusters that showed more activity for the looming than the static cue are located in posterior brain regions. These regions are posterior cingulate cortex (PCC; Brodmann area (BA) 23), extrastriate cortex (BA 19), and right precuneus (BA 7). The dipole cluster localized in the PCC consisted of 45 ICs from 28 unique subjects. The permutation test on the mean amplitude showed that there was a significant difference in processing the visual stimulus cued by a looming versus static sound ($F(1,44) = 4.40, p = .03$). The dipole cluster localized in the extrastriate cortex consisted of 84 ICs from 38 unique subjects. The permutation test on the mean amplitude showed that there was a significant difference in processing the visual stimulus cued by a looming versus static sound ($F(1,83) = 6.02, p = .02$). The dipole cluster localized in the precuneus consisted

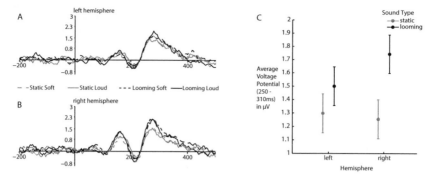

Figure 17: The ERP component for the looming cue benefit: rP280. The two waveforms on the left demonstrate the cue benefit in processing a target after subtracting auditory and visual processing (AV-AX-XV) for electrodes on the left hemisphere (A) and right hemisphere (B). C shows the statistical analysis of the average voltage amplitude (280 ±30 ms) for static and looming auditory cues in the left and right hemisphere. The results show a general looming cue benefit over the static cue that is more pronounced/driven by increased activity in the right hemisphere after a looming compared to a static auditory cue. The error bars display the 95% confidence intervals of the *Sound Type* x *Hemisphere* interaction effect.

of 67 ICs from 38 unique subjects. The permutation test on the mean amplitude showed that there was a significant difference in processing the visual stimulus cued by a looming versus static sound ($F(1,66) = 20.40, p < .001$).

We found the opposite, namely less activity for looming than static cues, in three dipole clusters located in frontal brain regions. These regions are right medial frontal cortex (MFC; BA 8), right inferior frontal cortex (rIFC; BA 45), and left medial frontal cortex (BA 10). The dipole cluster localized in the right MFC consisted of 82 ICs from 37 unique subjects. The permutation test on the mean amplitude showed that there was a significant difference in processing the visual stimulus cued by a looming versus static sound ($F(1,81) = 5.15, p = .02$). The dipole cluster localized in the rIFC consisted of 44 ICs from 29 unique subjects. The permutation test on the mean amplitude showed that there was a significant difference in processing the visual stimulus cued by a looming versus static sound ($F(1,43) = 4.14, p = .05$). The dipole cluster localized in the left MFC consisted of 36 ICs from 27 unique subjects. The permutation test on the mean amplitude showed that there was a significant difference in processing the visual stimulus cued by a looming versus static sound ($F(1,35) = 6.37, p = .01$).

To summarize, we identified an ERP component for the looming cue benefit, rP280. This looming cue benefit was stronger for the right posterior electrodes. Source estimation analysis identified two networks that might have given rise to this rP280 looming benefit. A posterior network consisting of PCC, extrastriate cortex, and precuneus that showed more activity for looming cues. In addition, there was activity in a frontal network consisting of bilateral MFC and rIFC that was less active for looming than static in the same time interval (280 ± 30 ms).

5.5 DISCUSSION

The current study shows that auditory *looming* cues bring about faster correct discrimination of peripheral visual targets than auditory *static* cues. In other words, we found a persistent benefit of looming sounds in cuing crossmodal attention reorienting. Louder cues can also speed up correct discriminations, but this benefit diminishes with time.

Previous studies have demonstrated preferential processing of looming sounds, in terms of behavior (Leo et al., 2011; Schouten et al., 2011; Sutherland et al., 2014; Cecere et al., 2014) as well as neural activity (Maier and Ghazanfar, 2007; Romei et al., 2009). To the best of our knowledge, we are the first to have identified an ERP correlate (rP280) for a looming benefit, specifically related to spatial attention cuing. To examine the cue benefit itself, we assumed an auditory and visual processing additivity. Subtracting activity caused by auditory or visual processing from the auditory-cued visual target processing leaves activity that can only be explained by the cue-target interaction. This revealed a right-hemispheric positive component, 280 ms after visual target onset, which was selectively larger for looming cues. Source estimation analysis identified two coexisting networks of neural regions that were likely to have given rise to the looming cue benefit. The posterior network was more active for looming than static cues which might have been propelled by decreased inhibition due to less frontal network activity for looming than static cues. We discuss these networks and their neural regions in more detail in the following sub-sections.

5.5.1 *Right hemispheric bias in ERP marker*

Auditory looming cues selectively induced a larger positive deflection at 280 ms (rP280) after a peripheral visual target appeared. The key characteristic of the rP280 is its right hemispheric lateralization in posterior electrodes. What might

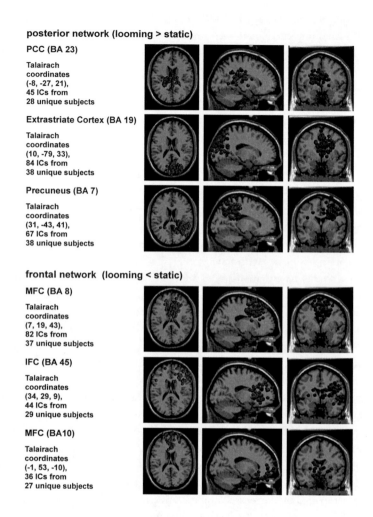

posterior network (looming > static)

PCC (BA 23)

Talairach
coordinates
(-8, -27, 21),
45 ICs from
28 unique subjects

Extrastriate Cortex (BA 19)

Talairach
coordinates
(10, -79, 33),
84 ICs from
38 unique subjects

Precuneus (BA 7)

Talairach
coordinates
(31, -43, 41),
67 ICs from
38 unique subjects

frontal network (looming < static)

MFC (BA 8)

Talairach
coordinates
(7, 19, 43),
82 ICs from
37 unique subjects

IFC (BA 45)

Talairach
coordinates
(34, 29, 9),
44 ICs from
29 unique subjects

MFC (BA10)

Talairach
coordinates
(-1, 53, -10),
36 ICs from
27 unique subjects

Figure 18: Brain regions active for rP280. Source estimation analysis identified 6 component cluster for 40 (2x20) subjects, grouped into a posterior and
a frontal network of neural regions. The posterior network shows significantly
higher activity for looming than static cues in PCC, extrastriate cortex, and right
precuneus. In contrast the frontal network shows significantly less activity for
looming than static cues in bilateral MFC and right IFC. Each blue dot represents
a single IC, the red dot is the average position of all blue dots belonging to this
dipole cluster.

give rise to this bias? Broadly speaking, the right parietal cortex is believed to underlie spatial awareness and is implicated in tasks that involve orienting spatial attention (Sturm and Willmes, 2001; Farah et al., 1989; Gitelman et al., 1999), localizing objects in space (Bushara et al., 1999; Zatorre et al., 2002; Sestieri et al., 2006; Arnott and Alain, 2011), and motion processing (Griffiths et al., 1998; Baumgart et al., 1999; Griffiths et al., 2000; Battelli et al., 2007). In contrast, the left parietal cortex is typically involved in linguistic processing (Geschwind, 1970; Cantalupo and Hopkins, 2001; Brownsett and Wise, 2010).

Consistent with this functional separation of the parietal lobe regions, looming sounds that signal an approaching object's location have been found to activate the right temporo-parietal junction (TPJ; Seifritz et al., 2002; Neuhoff et al., 2002; Bach et al., 2015). This preferential activity in the right parietal cortex is likely to reflect neural processing of the spatial information inherent to looming sounds. However, no previous work has investigated the neural activity underlying the looming cue benefit. Here, we show that the looming cue benefit (rP280) is also observed in right parietal areas. This lateralization underlying reorienting of spatial attention might be related to the spatial information processing fundamental to the parietal lobe.

5.5.2 *The role of auditory looming cues in reorienting spatial attention*

Source estimation for the looming cue benefit showed preferential activation of frontal and parietal areas that underlie the reorienting of attention. Previous work has shown two distinct fronto-parietal (FP) networks for reorienting attention. A dorsal FP network that controls attention voluntarily in a goal-driven manner and a ventral FP network that is activated involuntarily through behaviorally relevant stimuli (Corbetta and Shulman, 2002; Corbetta et al., 2008; Corbetta and Shulman, 2011). This functional differentiation of the two networks is also apparent from the involvement of brain regions. While the dorsal attention system recruits parietal areas such as the superior parietal lobe (SPL) and inferior parietal sulcus (IPS), the ventral system typically involves the TPJ. Nevertheless, the recruitment of frontal areas for the two networks overlaps, which enables the necessary interaction of the dorsal and ventral system to reorient spatial attention.

Previous work reported that looming sounds preferentially activate a parietal area that belongs to the ventral FP attention network, namely the TPJ (Seifritz et al., 2002; Neuhoff et al., 2002; Bach et al., 2008, 2015). Unlike this work, we did not find increased TPJ activity given that we investigated the sustained cuing response to looming sounds and not the looming sound itself. The enhanced TPJ activity for processing looming sounds might reflect the increased capture of attention by looming sounds prior to preferential visual processing of the attended location. Through the connecting frontal regions, the location of relevance might be communicated from the ventral system, that detected the stimulus, to the dorsal system, which is responsible to shift attention voluntarily.

Our results demonstrate that looming cues preferentially activate the dorsal attention network. The increased SPL activity for the looming cue benefit might reflect the enhanced voluntary reorienting of spatial attention to visual targets. More precisely, the estimated dipole's localization lies in the precuneus which is believed to take part in the voluntary orienting of spatial attention (Culham et al., 1998; Le et al., 1998; Simon et al., 2002; Cavanna and Trimble, 2006). The activity in the dorsal FP network for orienting attention voluntarily can also extend to visual cortices. Previous work has shown that the activity from the dorsal FP network also increases activity in the extrastriate cortex (Kastner et al., 1999; Hopfinger et al., 2000). Thus, our increased extrastriate activation for looming cues might reflect the top-down bias to preferentially sustain attention to the cued location after a looming sound, in spite of the demands of the central task. In other words, the increased activity for posterior areas of the dorsal attention network could reflect the increased urge to reorient attention voluntarily away from the central visuo-motor tracking task to perform the peripheral discrimination task.

The preferential deactivation of frontal areas for the looming cue benefit supports the (stronger) reorienting of spatial attention. The prefrontal cortex is generally known to be involved in executive functions such as controlling the reorienting of attention to switch tasks (Miller and Cohen, 2001). In this context, the MFC plays a major role in transferring information (i.e., location of relevant information) from the ventral system to the dorsal system in order to know where attention should be reorient towards (Fox et al., 2006). Our source estimation analysis showed that the bilateral MFC (BA 8 and BA 10) but also the rIFC (BA 45) contributed to the looming cue benefit. The rIFC is believed to control the reorienting of attention by inhibiting the reorientation of attention for irrelevant information (see Aron et al., 2004, 2014, for reviews). Thus, less activity for looming cues suggests less inhibition to shift attention. In other words, looming cues increased attention shifting compared to static cues.

Additionally, decreased frontal activity for auditory looming cues may also support the maintenance of attention at the reoriented location. The FEF, which is part of the right MFC, is thought to produce inhibition of return (IOR; Ro et al., 2003). IOR removes attention from an attended location and inhibits the reorienting to this previously attended location. We found less right MFC activity, thus less FEF activity for the looming than for the static cue benefit. The decreased FEF activity suggests less production of IOR for looming cued targets. This activity pattern might explain why the exogenous looming cue, unlike other auditory cues, allowed sustained attention until the sound offset in a previous study (Glatz and Chuang, 2018, under review).

Besides the cortical areas that identified with regions from the FP networks, we also found preferential activity in the PCC, a subcortical region that is part of the limbic system. Two potential reasons can account for the preferential activation of the PCC for the auditory looming cue benefit. First, it could reflect the increased emotional processing (Maddock et al., 2003) due to the attributed threat to looming stimuli. In line with this, previous work found preferential activation of the amygdala for looming sounds (Bach et al., 2008). The alternative, and preferred, explanation is that PCC activity influences attentional focus. In task-switching paradigms, such as our experiment, increased PCC activity is thought to reflect the ability to quickly change from one cognitive state to another (Leech and Sharp, 2014). Hence, the preferential PCC activity related to the looming cue benefit could reflect the increased readiness to shift attention away from the visuo-motor tracking task to the discrimination task.

In summary, our results support the activation of the FP networks for shifting spatial attention. In addition, we were able to show that these networks responded preferentially to looming cues. The top-down control of frontal regions on parietal regions is in favor of the looming cue. It could allow the ventral network to detect the relevant looming cue better which favors the subsequent voluntary orienting of attention for looming cues despite the required central focus and might sustain this attention as long as necessary, that is, until the sound offset.

5.5.3 *Why do looming cues favor spatial attention reorienting?*

Auditory looming cues provide information about object motion in egocentric space. When objects enter the immediate space surrounding the body, also known as peripersonal space (PPS), they can pose a potential threat to the body. Hence, approaching sounds that signal these objects receive more attention when entering the PPS than when approaching at further distance (Canzoneri et al., 2012). To preserve a margin of safety around the body, the PPS boundaries are extended for looming objects (Noel et al., 2015; Ferri et al., 2015). This means that objects enter the PPS earlier and, therefore, receive attention earlier. This sensitivity to looming objects gives individuals additional time to prepare for the object's arrival. In contrast, if the PPS boundaries were closer to the body, approaching objects would cross the margin of safety later and, therefore, afford less time to prepare for the object's arrival. This shaping of the PPS boundaries is presumably achieved by preferential activation of intraparietal cortices, which represent the PPS, for looming over receding objects (Graziano and Cooke, 2006; Clery et al., 2015; di Pellegrino and Làdavas, 2015).

Taken together, the virtue of auditory looming cues in reorienting visuo-spatial attention crossmodally might be due to its ability to continuously communicate spatial information. An object's location in space can be extracted from an approaching sound and, subsequently, be integrated into a supramodal space representation. Visuo-spatial attention can benefit from this internal map by shifting attention to the location of interest, specified by the auditory looming cue. The preferential activation of supramodal parietal areas for looming might reflect the increased spatial information processing of relevant locations. In combination with the reduced frontal activity for the looming cue benefit, it might reflect the enhanced spatial attention reorienting. This can be observed in the improved behavioral performance (i.e., faster RTs) for looming cued targets.

5.5.4 Future directions

To date, surprisingly little work has been performed on the utility of auditory looming cues in shifting visuo-spatial attention. Hence, many questions remain to be answered in this new area of research. Future studies could, for example, investigate the influence of different cue-target onset asynchronies (CTOA) between auditory looming cues and visual targets, different looming cue durations, or manipulate the velocity of looming cues. Here, we determined the CTOA to exclude any multisensory effect by presenting the auditory cue and visual target separately without any overlap. Nevertheless, future studies can consider the use of multisensory looming cues to investigate how multisensory stimulation interacts with the looming cuing benefit.

Given the limitations of source estimation, future research on looming cues should employ neuroimaging techniques with a higher spatial resolution (e.g., fMRI) to validate the networks of regions identified here. This work might also investigate the connectivity between these regions. Based on previously shown fronto-parietal attention networks (Corbetta and Shulman, 2002; Thiel et al., 2004; Gitelman et al., 1999), the inhibitory frontal network we identified might interact with the posterior network. More specifically, previous work proposes the PCC as mediator between the frontal and parietal areas (Leech and Sharp, 2014). In this role, the PCC receives input from the prefrontal cortex that determines its level of activity. This PCC activity was found to correlated positively with activations of parietal cortices involved in spatial attention (Mohanty et al., 2008).

5.6 CONCLUSION

To the best of our knowledge, this is the first work that demonstrates an electrophysiological marker, rP280, for the auditory looming cue benefit in reorienting visuo-spatial attention to peripheral tasks. This looming cue benefit is also observed behaviorally, in faster peripheral target discrimination. In addition, we identified two networks of neural regions that gave rise to the looming cue benefit. One prominent region from the posterior network is the parietal lobe where spatial attention is administered. Spatial information, that is an inherent property of looming sounds, is most likely integrated crossmodally in parietal regions. The coexistent frontal network might further amplify this selective activation by auditory looming through reduced inhibition for shifting attention. In summary, auditory looming cues might be especially efficient in reorienting visuo-spatial attention due to their valuable spatial information content in constructing an environmental representation relevant to the body.

6

APPENDIX: THE PERSISTENCE OF THE AUDITORY
LOOMING CUE BENEFIT POST CUE PRESENTATION

6.1 INTRODUCTION

Chapter 4 demonstrated that the auditory looming cue benefit can be observed after the approaching auditory motion has unfolded, at the sound offset but not earlier (i.e., CTOA 0 ms and 250 ms). This suggests that auditory looming cues function as an exogenous cues but also exert an endogenous influence on visuo-spatial attention. If the auditory looming cue indeed reflects some endogenous properties, how long can spatial attention be voluntarily sustained at the cued location? To investigate this question, the gap between sound and target can be manipulated.

Previous work employing auditory looming cues suggest increased alertness even after the sound offset. Bach et al. (2009) introduced a gap of 100 ms between the auditory cue offset and target onset and were still able to observe increased alertness for looming sounds. In a subsequent experiment, Bach et al. (2015) investigated the time-course of visual phasic alertness more specifically by introducing 200, 400, and 600 ms gaps between the cue offset and the visual target onset. Interestingly, their results revealed a looming cue benefit for visual targets presented 400 ms after the sound offset but not 200 ms earlier or later (i.e., 200 ms or 600 ms gap). Taken together, the results of these two studies suggest that a looming cue benefit can continue to exist after the sound offset even though the exact time-course is unknown.

The above studies demonstrated increased phasic alertness after a gap that resulted in cue-target benefits. However, the introduction of a concurrent task that competes for attentional resources (see Chapter 1.3.4) could negate such a cue benefit. In a dual-task paradigm, sustaining attention to one task comes at the cost of the other task's performance. My previous studies (Chapter 4 and 5) demonstrated an auditory looming cue benefit in shifting spatial attention immediately after its offset during dual-tasking. Here, I examined whether this benefit continues to persist at 500 ms after cue offset. This was evaluated based on performance measures (d', RT, RMSE) and neural responses (ERP) to the visual target.

If the following sections do not state otherwise, the methods used for this study were the same as reported for the study in Chapter 5.3.

6.2.1 Participants

Twenty healthy volunteers (mean age = 25.45 $\pm$ 3.15; 8 males) participated in the current experiment. All participants reported normal (or corrected-to-normal) vision, normal hearing, and no known history of neurological problems. All participants received written instructions, gave their signed consent, and were remunerated 12 Euros/hour for their voluntary participation. The experimental procedure was approved by the ethics council of the University Hospital Tübingen.

6.2.2 Stimuli

In this experiment I used different auditory cues than in Chapter 5.3.2 to precede the visual target. The auditory cues were also 400 Hz triangular waveforms, sampled at 44.1 kHz, of 500 ms duration. However, their intensity profile was either looming, receding (the inverse of the looming), or static. I used the same looming sound as in Chapter 5.3.2 (loud, $30 - 75$ dB), a receding sound to control for the profile dynamics ($75 - 30$ dB), and a static sound (65 dB) to control for the average intensity of these two sounds. Sound pressure levels were measured at participants' approximate head position. All sounds had 5 ms ramps at the on- and offset to avoid clicking noise at the beginning and end of sound presentation. These ramps were achieved by convolving the sound with a trapezoid grating.

6.2.3 Procedure

In contrast to Chapter 5.3.4, this study consisted of one session only. This session contained 20 experimental blocks (4.5 min per block) that were separated by mandatory 1.5 min breaks.

Again, participants were asked to continuously perform the central visuo-motor tracking task and occasionally discriminate peripheral visual targets. While the visual target immediately followed the auditory cue in the AV condition in

Chapter 5.3.4 (CTOA 500 ms), it now appeared 500 ms after the auditory cue offset such that the visual target was presented at a CTOA of 1000 ms.

6.3 RESULTS AND DISCUSSION

6.3.1 Behavioral performance

Participants' performance in the peripheral tilt discrimination task was assessed by calculating the discrimination sensitivity (d') and the correctly identified response times (RT) to visual targets. The discrimination sensitivity for the visual targets was calculated through the hits and false alarms for the different conditions (Macmillan and Creelman, 1991).

The performance in the continuous tracking task was assessed as the root-mean-squared-error (RMSE) by calculating the deviation between the cursor's position and the designated center during the 250 ms visual target presentation.

INFLUENCE OF CUING   First, I analyzed whether cuing a visual target influences the performance for the three performance measures (cue vs. no-cue). I found a significant difference for d' (1-tailed, paired-sample t-test; $t(19) = -2.94, p < .01, d = -.66$), as well as for RT (1-tailed, paired-sample t-test; $t(19) = 5.06, p < .001, d = 1.13$), but not for RMSE (1-tailed, paired-sample t-test; $t(19) = 1.08, p = .15, d = .24$).

This suggest that auditory cues, regardless of sound type, can benefit the secondary task by improving target discrimination and, at the same time, speed responses to the visual targets. The smaller error in the tracking performance during cued trials was not statistically significant.

PERFORMANCE ON CUED TRIALS   In a second step, I submitted the respective measurement data (d', RT, RMSE) to a repeated measurements analysis of variance (ANOVA) with the factor *Sound Type* (static, receding, looming).

There was no effect of *Sound Type* for any of the three performance measures; d' ($F(2,38) = 3.14, p = .06, \omega^2 = .10$), RT ($F(2,38) = .44, p = .65, \omega^2 = .00$), and RMSE ($F(2,38) = .41, p = .67, \omega^2 = .00$).

This suggests that while auditory cues still improve performance for the visual targets even after introducing a 500 ms gap at the sound offset, the type of sound used to cue the visual target does not influence the response accuracy or

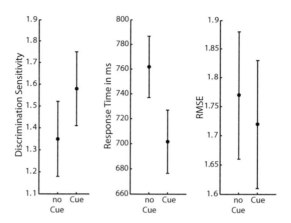

Figure 19: Behavioral performance to visual targets that could be cued by a sound or not. The left shows the significantly better discrimination sensitivity when cued compared to no-cued targets. The middle shows the significantly faster response times when the visual target was preceded by an auditory cues. The right shows the tracking performance during the sound. Error bars display the 95% confidence intervals of the effect.

speed. That is, the looming benefit that was observed in faster response times at the sound offset did not hold until 500 ms later.

ERP RESULTS   Even though I did not observe a behavioral benefit for looming cued visual targets, the brain might still respond preferentially to them. To investigate whether the auditory looming cue benefit in visual processing was still present 500 ms later, I subtracted, as in Chapter 5.4.2, the activity related to auditory processing (AX) and the activity related to uncued visual target processing (XV) from the activity associated with the combined cued visual processing (AV-AX-XV). Therefore, the remaining activity reflects the advantage in visual processing of peripheral targets due to an auditory cue.

The ERP waveforms for the auditory cue benefit according to the three different sounds are displayed in Figure 20. These results are in congruence with the behavioral findings. That is, while there is still an auditory cue benefit as reflected at approximately 820 ms, this benefit does not differ across auditory cue types. Hence, looming sounds did not cue attention better than comparable sounds after introducing a 500 ms gap which was also mirrored in not significantly different response times.

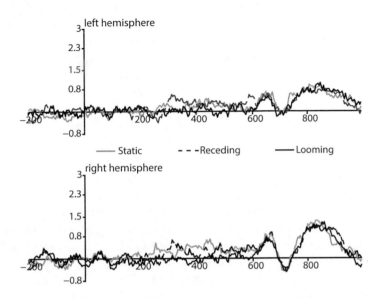

Figure 20: The two waveforms demonstrate the auditory cue benefit at posterior electordes in processing a target that appeared 500 ms after the sound ended. For this, auditory and visual processing was subtracted from the audio-visual processing (AV-AX-XV) for electrodes on the left and right posterior hemispheres. After introducing the 500 ms gap between sound offset and visual onset, no sound-specific cue benefit can be observed at 820 ms after the sound offset.

Nevertheless, a hemispheric difference for the spatial cue benefit through sounds is observable. As in Chapter 5.4.2, the cue benefit is more pronounced on the right than the left hemisphere. I could have found stronger activation of the right posterior hemisphere (i.e., right parietal brain regions) because the current task required shifting of spatial attention; this is believed to originate from right parietal regions (Sturm and Willmes, 2001; Farah et al., 1989; Gitelman et al., 1999).

These results suggest that the auditory looming cue captures transient attention rather than voluntarily sustaining attention at the cued location. If the looming sound would function as an endogenous cue, one would expect the participants to sustain attention at the cued location. This would then, even after a 500 ms gap, result in faster responses, especially since participants were aware of the 500 ms gap.

Alternatively, it could be that the implicit meaning of the looming sound, namely of an approaching object, precludes sustaining attention longer than the duration of the sound itself. When a looming sound stops, it can either indicate that the object has arrived or that it is no longer relevant. The results by Burton (2011) are in agreement. He demonstrated that task-irrelevant looming sounds were able to facilitate movements in response to visual targets. His experiments showed, in congruence with the findings of Chapter 5.4 and Chapter 4.3.1, that the looming benefit was largest at the sound offset. Like here, Burton (2011) did not observe the movement facilitation for looming sounds anymore when introducing a 500 ms gap after the 1000 ms looming sound (i.e., CTOA 1500 ms).

Taken together, the findings by Burton (2011), Chapter 4, and this experiment suggest that looming sounds are especially useful at the sound offset when they have conveyed all their information, that is when the object has arrived. In this regard, the duration of the looming sound might not necessarily effect the looming benefit. Romei et al. (2009) found increased visual cortex excitability at the end of looming sounds independent of their duration (250, 500, 1000 ms). Nevertheless, future work might investigate if and how the duration of an auditory looming cue influences the response to (peripheral visual) targets.

Adams, R. B. and Janata, P. (2002). A comparison of neural circuits underlying auditory and visual object categorization. *NeuroImage*, 16(2):361–377.

Adcock, M. and Barrass, S. (2004). Cultivating design patterns for auditory displays. In *Proceedings of ICAD 04-Tenth Meeting of the International Conference on Auditory Display*, pages 4–7.

Alain, C. (2007). Breaking the wave: Effects of attention and learning on concurrent sound perception. *Hearing Research*, 229(1-2):225–236.

Arnott, S. R. and Alain, C. (2011). The auditory dorsal pathway: Orienting vision. *Neuroscience and Biobehavioral Reviews*, 35(10):2162–2173.

Aron, A. R., Robbins, T. W., and Poldrack, R. A. (2004). Inhibition and the right inferior frontal cortex. *Trends in Cognitive Sciences*, 8(4):170–177.

Aron, A. R., Robbins, T. W., and Poldrack, R. A. (2014). Inhibition and the right inferior frontal cortex: One decade on. *Trends in Cognitive Sciences*, 18(4):177–185.

Ashmead, D. H., Davis, D. L., and Northington, A. (1995). Contribution of listeners' approaching motion to auditory distance perception. *Journal of Experimental Psychology: Human Perception and Performance*, 21(2):239–256.

Bach, D. R., Furl, N., Barnes, G., and Dolan, R. J. (2015). Sustained magnetic responses in temporal cortex reflect instantaneous significance of approaching and receding sounds. *PLoS ONE*, 10(7):7–9.

Bach, D. R., Neuhoff, J. G., Perrig, W., and Seifritz, E. (2009). Looming sounds as warning signals: The function of motion cues. *International Journal of Psychophysiology*, 74(1):28–33.

Bach, D. R., Schächinger, H., Neuhoff, J. G., Esposito, F., Di Salle, F., Lehmann, C., Herdener, M., Scheffler, K., and Seifritz, E. (2008). Rising sound intensity: An intrinsic warning cue activating the amygdala. *Cerebral Cortex*, 18(1):145–50.

Bainbridge, L. (1983). Ironies of automation. *Automatica*, 19(6):775–779.

Ball, W. and Tronick, E. (1971). Infant responses to impending collision: Optical and real. *Science*, 171:818–820.

Ballas, J. A. (1993). Common factors in the identification of an assortment of brief everyday sounds. *Journal of Experimental Psychology: Human Perception and Performance*, 19(2):250–267.

Ballas, J. A. and Howard, J. H. (1987). Interpreting the language of environmental sounds. *Environment and behavior*, 19(1):91–114.

Battelli, L., Pascual-Leone, A., and Cavanagh, P. (2007). The 'when' pathway of the right parietal lobe. *Trends in Cognitive Sciences*, 11(5):204–210.

Baumgart, F., Gaschler-Markefski, B., Woldorff, M. G., Heinze, H.-J., and Scheich, H. (1999). A movement-sensitive area in auditory cortex. *Nature*, 400:724–726.

Baumgartner, R., Reed, D. K., Tóth, B., Best, V., Majdak, P., Colburn, H. S., and Shinn-Cunningham, B. (2017). Asymmetries in behavioral and neural responses to spectral cues demonstrate the generality of auditory looming bias. *Proceedings of the National Academy of Sciences of the United States of America*, pages 1–6.

Belz, S. M., Robinson, G. S., and Casali, J. G. (1999). A new class of auditory warning signals for complex systems: Auditory icons. *Human Factors*, 41(4):608–618.

Benjamini, Y. and Hochberg, Y. (1995). Controlling the false discovery rate: A practical and powerful approach to multiple testing. *Journal of the Royal Statistical Society*, 57(1):289–300.

Bentin, S., McCarthy, G., and Wood, C. C. (1985). Event-related potentials, lexical decision and semantic priming. *Electroencephalography and Clinical Neurophysiology*, 60(4):343–355.

Bigdely-Shamlo, N., Mullen, T., Kothe, C., Su, K.-M., and Robbins, K. A. (2015). The PREP pipeline: Standardized preprocessing for large-scale EEG analysis. *Frontiers in Neuroinformatics*, 9:1–20.

Bleichner, M. G., Mirkovic, B., and Debener, S. (2016). Identifying auditory attention with ear-EEG: cEEGrid versus high-density cap-EEG comparison. *Journal of Neural Engineering*, 13:1–13.

Borojeni, S. S., Chuang, L., Heuten, W., and Boll, S. (2016). Assisting drivers with ambient take-over requests in highly automated driving. *Proceedings of the 8th International Conference on Automotive User Interfaces and Interactive Vehicular Applications - Automotive'UI 16*, pages 237–244.

Brainard, D. H. (1997). The psychophysics toolbox. *Spatial vision*, 10(4):433–436.

Bregman, A. S. (1990). The perceptual organization of sound. In *Foundations of Cognitive Psychology*, pages 213–248.

Brewer, M. B. (2000). Research design and issues of validity. In *Handbook of Research Methods in Social and Personality Psychology*, pages 3–16.

Broadbent, D. E. (1957). A mechanical model for human attention and immediate memory. *Psychol Rev*, 64(3):205–215.

Broadbent, D. E. (1971). *Decision and Stress*.

Brouwer, A. M., Hogervorst, M. A., Van Erp, J. B., Heffelaar, T., Zimmerman, P. H., and Oostenveld, R. (2012). Estimating workload using EEG spectral power and ERPs in the n-back task. *Journal of Neural Engineering*, 9(4).

Brownsett, S. L. and Wise, R. J. (2010). The contribution of the parietal lobes to speaking and writing. *Cerebral Cortex*, 20(3):517–523.

Burton, J. (2011). *Linkages between auditory perception and action: Acoustical facilitation of motor responses*. PhD thesis.

Bushara, K. O., Weeks, R. A., Ishii, K., Catalan, M.-J., Tian, B., Rauschecker, J. P., and Hallett, M. (1999). Modality-specific frontal and parietal areas for auditory and visual spatial localization in humans. *Nature Neuroscience*, 2(8):759–766.

Cantalupo, C. and Hopkins, W. D. (2001). Asymmetric Broca's area in great apes. *Nature*, 414(6863):505–505.

Canzoneri, E., Magosso, E., and Serino, A. (2012). Dynamic sounds capture the boundaries of peripersonal space representation in humans. *PLoS ONE*, 7(9):3–10.

Cao, Y., van der Sluis, F., Theune, M., op den Akker, R., and Nijholt, A. (2010). Evaluating informative auditory and tactile cues for in-vehicle information systems. In *Proceedings of the 2nd International Conference on Automotive User Interfaces and Interactive Vehicular Applications - AutomotiveUI '10*, pages 102–

109.

Cappe, C., Thelen, A., Romei, V., Thut, G., and Murray, M. M. (2012). Looming signals reveal synergistic principles of multisensory integration. *The Journal of Neuroscience*, 32(4):1171–1182.

Cappe, C., Thut, G., Romei, V., and Murray, M. M. (2009). Selective integration of auditory-visual looming cues by humans. *Neuropsychologia*, 47(4):1045–52.

Carrasco, M. (2011). Visual attention: The past 25 years. *Vision Research*, 51(13):1484–1525.

Casner, S. M., Hutchins, E. L., and Norman, D. (2016). The challenges of partially automated driving. *Communications of the ACM*, 59(5):70–77.

Cavanna, A. E. and Trimble, M. R. (2006). The precuneus: A review of its functional anatomy and behavioural correlates. *Brain*, 129(3):564–583.

Cecere, R., Romei, V., Bertini, C., and Làdavas, E. (2014). Crossmodal enhancement of visual orientation discrimination by looming sounds requires functional activation of primary visual areas: A case study. *Neuropsychologia*, 56:350–358.

Cheal, M. L., Lyon, D. R., and Hubbard, D. C. (1991). Does attention have different effects on line orientation and line arrangement discrimination? *The Quarterly Journal of Experimental Psychology Section A*, 43(4):825–857.

Cherry, E. C. (1953). Some experiments on the recognition of speech, with one and with two ears. *The Journal of the Acoustical Society of America*, 25(5):975–979.

Chuang, L. L., Glatz, C., and Krupenia, S. (2017). Using EEG to understand why behavior to auditory in-vehicle notifications differs across test environments. *Proceedings of the 9th International Conference on Automotive User Interfaces and Interactive Vehicular Applications*, pages 123–133.

Chun, M. M., Golomb, J. D., and Turk-Browne, N. B. (2011). A taxonomy of external and internal attention. *Annual Review of Psychology*, 62(1):73–101.

Clery, J., Guipponi, O., Odouard, S., Wardak, C., and Ben Hamed, S. (2015). Impact prediction by looming visual stimuli enhances tactile detection. *Journal of Neuroscience*, 35(10):4179–4189.

Comerchero, M. D. and Polich, J. (1999). P3a and P3b from typical auditory and visual stimuli. *Clinical Neurophysiology*, 110(1):24–30.

Corbetta, M., Patel, G., and Shulman, G. L. (2008). The reorienting system of the human brain: From environment to theory of mind. *Neuron*, 58(3):306–324.

Corbetta, M. and Shulman, G. L. (2002). Control of goal-directed and stimulus-driven attention in the brain. *Nature Reviews Neuroscience*, 3(3):201–215.

Corbetta, M. and Shulman, G. L. (2011). Spatial neglect and attention networks. *Annual Review of Neuroscience*, 34(1):569–599.

Crowley, K. E. and Colrain, I. M. (2004). A review of the evidence for P2 being an independent component process: Age, sleep and modality. *Clinical Neurophysiology*, 115(4):732–744.

Culham, J. C., Brandt, S. A., Cavanagh, P., Kanwisher, N. G., Dale, A. M., and Tootell, R. B. (1998). Cortical fMRI activation produced by attentive tracking of moving targets. *Journal of neurophysiology*, 80(5):2657–2670.

Cummings, A., Čeponiene, R., Koyama, A., Saygin, A. P., Townsend, J., and Dick, F. (2006). Auditory semantic networks for words and natural sounds. *Brain*

Research, 1115(1):92–107.

Davidson, M. C. and Marrocco, R. T. (2000). Local infusion of scopolamine into intraparietal cortex slows covert orienting in rhesus monkeys. *Journal of Neurophysiology*, 83:1536–1549.

Dehais, F., Causse, M., Vachon, F., Régis, N., Menant, E., and Tremblay, S. (2014). Failure to detect critical auditory alerts in the cockpit: Evidence for inattentional deafness. *Human Factors*, 56(4):631–644.

Delacre, M., Lakens, D., and Leys, C. (2017). Why psychologists should by default use Welch's t -test instead of Student's t -test. *International Review of Social Psychology*, 30(1):92.

Delorme, A. and Makeig, S. (2004). EEGLAB: an open sorce toolbox for analysis of single-trail EEG dynamics including independent component anlaysis. *Journal of Neuroscience Methods*, 134:9–21.

Delorme, A., Palmer, J., Onton, J., Oostenveld, R., and Makeig, S. (2012). Independent EEG sources are dipolar. *PLoS ONE*, 7(2).

di Pellegrino, G. and Làdavas, E. (2015). Peripersonal space in the brain. *Neuropsychologia*, 66:126–133.

Dick, F., Bussiere, J., and Saygin, A. (2002). The effects of linguistic mediation on the identification of environmental sounds. *Center for Research in Language Newsletter*, 14(3):3–9.

Dick, F., Saygin, A. P., Galati, G., Pitzalis, S., Bentrovato, S., D'Amico, S., Wilson, S., Bates, E., and Pizzamiglio, L. (2007). What is involved and what is necessary for complex linguistic and nonlinguistic auditory processing: Evidence from functional magnetic resonance imaging and lesion data. *Journal of Cognitive Neuroscience*, 19(5):799–816.

Dien, J., Spencer, K. M., and Donchin, E. (2004). Parsing the late positive complex: Mental chronometry and the ERP components that inhabit the neighborhood of the P300. *Psychophysiology*, 41(5):665–678.

Dingler, T., Lindsay, J., and Walker, B. N. (2008). Learnability of sound cues for environmental features: Auditory icons, earcons, spearcons, and speech. In *14th International Conference on Auditory Display*, pages 1–6.

Donchin, E. (1981). Surprise! ... Surprise? *Pyscholohysiology*, 18(5):493–513.

Donchin, E. and Coles, M. G. H. (1988). Is the P300 component a manifestation of context updating? *Behavioral and Brain Sciences*, 11(3):355–425.

Donohue, S. E., Green, J. J., and Woldorff, M. G. (2015). The effects of attention on the temporal integration of multisensory stimuli. *Frontiers in Integrative Neuroscience*, 9:32.

Dosenbach, N. U., Fair, D. A., Cohen, A. L., Schlaggar, B. L., and Petersen, S. E. (2008). A dual-networks architecture of top-down control. *Trends in Cognitive Sciences*, 12(3):99–105.

Driver, J. and Spence, C. (1998). Crossmodal attention. *Current Opinion in Neurobiology*, 8:245–253.

Duncan-Johnson, C. C. and Donchin, E. (1977). On quantifying surprise: The variation of event-related potentials with subjective probability. *Psychophysiology*, 14(5):456–467.

Edman, T. R. (1982). Human factors guidelines for the use of synthetic speech devices. In *Proceedings of the Human Factors and Ergonomics Society 26th Annual Meeting*, pages 212–216.

Edworthy, J. (1994). The design and implementation of non-verbal auditory warnings. *Applied Ergonomics*, 25(4):202–210.

Edworthy, J. and Hards, R. (1999). Learning auditory warnings: The effects of sound type, verbal labelling and imagery on the identification of alarm sounds. *International Journal of Industrial Ergonomics*, 24(6):603–618.

Edworthy, J. and Hellier, E. (2006). Alarms and human behaviour: Implications for medical alarms. *British Journal of Anaesthesia*, 97(1):12–17.

Edworthy, J., Loxley, S., and Dennis, I. (1991). Improving auditory warning design: Relationship between warning sound parameters and perceived urgency. *Human Factors: The Journal of the Human Factors and Ergonomics Society*, 33(2):205–231.

Egeth, H. E. and Yantis, S. (1997). Visual attention: Control, respresentation and time course. 48:269–297.

Egly, R. and Homa, D. (1991). Reallocation of visual attention. *Journal of Experimental Psychology: Human Perception and Performance*, 17(1):142–159.

Eimer, M. (1993). Spatial cueing, sensory gating and selective response preparation: An ERP study on visuo-spatial orienting. *Electroencephalography and Clinical Neurophysiology*, 88(5):408–420.

Eimer, M. (1994). "Sensory gating" as a mechanism for visuospatial orienting: Electrophysiological evidence from trial-by-trial cuing experiments. *Perception & Psychophysics*, 55(6):667–675.

Eimer, M. (2001). Crossmodal links in spatial attention between vision, audition, and touch: Evidence from event-related brain potentials. *Neuropsychologia*, 39(12):1292–1303.

Eimer, M. and Driver, J. (2001). Crossmodal links in endogenous and exogenous spatial attention: Evidence from event-related brain potential studies. *Neuroscience and Biobehavioral Reviews*, 25(6):497–511.

Eimer, M., Maravita, A., Van Velzen, J., Husain, M., and Driver, J. (2002). The electrophysiology of tactile extinction: ERP correlates of unconscious somatosensory processing. *Neuropsychologia*, 40(13):2438–47.

Eimer, M. and Schröger, E. (1998). ERP effects of intermodal attention and crossmodal links in spatial attention. *Psychophysiology*, 35(3):313–27.

Eimer, M., Van Velzen, J., Forster, B., and Driver, J. (2003). Shifts of attention in light and in darkness: An ERP study of supramodal attentional control and crossmodal links in spatial attention. *Cognitive Brain Research*, 15(3):308–323.

Engström, J., Markkula, G., Victor, T., and Merat, N. (2017). Effects of cognitive load on driving performance: The cognitive control hypothesis. *Human Factors: The Journal of the Human Factors and Ergonomics Society*, 59(5):734–764.

Eriksen, C. W. and Hoffman, J. E. (1972). Temporal and spatial characteristics of selective coding from visual displays. *Perception and Psychophysics*, 12(2B):201–204.

Ernst, M. O. and Bülthoff, H. H. (2004). Merging the senses into a robust percept. *Trends in Cognitive Sciences*, 8(4):162–169.

Escera, C., Alho, K., Schröger, E., and Winkler, I. W. (2000). Involuntary attention and distractibility as evaluated with event-related brain potentials. *Audiology and Neurotology*, 5(3-4):151–166.

Fabiani, M., Karis, D., and Donchin, E. (1986). P300 and recall in an incidental memory paradigm. *Psychophysiology*, 23(3):298–308.

Fabiani, M., Kazmerski, V. A., Cycowicz, Y. M., and Friedman, D. (1996). Naming norms for brief environmental sounds: Effects of age and dementia. *Psychophysiology*, 33:462–475.

Fagerlönn, J., Lindberg, S., and Sirkka, A. (2015). Combined auditory warnings for driving-related information. In *Proceedings of the Audio Mostly 2015 on Interaction With Sound*, pages 1–5.

Fagot, C. A. (1994). *Chronometric investigations of task switching.* PhD thesis.

Farah, M. J., Wong, A. B., Monheit, M. A., and Morrow, L. A. (1989). Parietal lobe mechanisms of spatial attention: Modality-specific or supramodal? *Neuropsychologia*, 27(4):461–470.

Ferri, F., Tajadura-Jiménez, A., Väljamäe, A., Vastano, R., and Costantini, M. (2015). Emotion-inducing approaching sounds shape the boundaries of multisensory peripersonal space. *Neuropsychologia*, 70:468–475.

Fox, M. D., Corbetta, M., Snyder, A. Z., Vincent, J. L., and Raichle, M. E. (2006). Spontaneous neuronal activity distinguishes human dorsal and ventral attention systems. *Proceedings of the National Academy of Sciences*, 103(26):10046–10051.

Franconeri, S. L. and Simons, D. J. (2003). Moving and looming stimuli capture attention. *Perception & Psychophysics*, 65(7):999–1010.

Friedman, D., Cycowicz, Y. M., and Gaeta, H. (2001). The novelty P3: An event-related brain potential (ERP) sign of the brain's evaluation of novelty. *Neuroscience and Biobehavioral Reviews*, 25(4):355–373.

García-Larrea, L., Lukaszewicz, A. C., and Mauguiére, F. (1992). Revisiting the oddball paradigm. Non-target vs neutral stimuli and the evaluation of ERP attentional effects. *Neuropsychologia*, 30(8):723–741.

Gaver, W. (1989). The SonicFinder: An interface that uses auditory icons. *Human-Computer Interaction*, 4(1):67–94.

Geschwind, N. (1970). The organization of language and the brain. *Science*, 170(3961):940–944.

Giordano, A. M., McElree, B., and Carrasco, M. (2009). On the automaticity and exibility of covert attention: A speed-accuracy trade-off analysis. *Journal of Vision*, 9(3)(30):1–10.

Gitelman, D. R., Nobre, A. C., Parrish, T. B., LaBar, K. S., Kim, Y. H., Meyer, J. R., and Mesulam, M. M. (1999). A large-scale distributed network for covert spatial attention. Further anatomical delineation based on stringent behavioural and cognitive controls. *Brain*, 122(6):1093–1106.

Glatz, C., Bülthoff, H. H., and Chuang, L. L. (2016). Why do auditory warnings during steering allow for faster visual target recognition? In *1st Neuroergonomics Conference: The Brain at Work and in Everyday Life, Neuroergonomics Conference 2016*.

Glatz, C. and Chuang, L. L. (2018). The time course of auditory looming cues in a divided visual attention paradigm.

Graham, R. (1999). Use of auditory icons as emergency warnings: Evaluation within a vehicle collision avoidance application. *Ergonomics*, 42(9):1233–48.

Graham, R., Hirst, S. J., and Carter, C. (1995). Auditory icons for collision-avoidance warnings. In *Intelligent Transportation: Serving the User Through Deployment. Proceedings of the 1995 Annual Meeting of ITS America*.

Gramann, K., Ferris, D. P., Gwin, J., and Makeig, S. (2014). Imaging natural cognition in action. *International Journal Psychophysiology*, 91(1):22–29.

Gray, R. (2011). Looming auditory collision warnings for driving. *Human Factors: The Journal of the Human Factors and Ergonomics Society*, 53(1):63–74.

Graziano, M. S. A. and Cooke, D. F. (2006). Parieto-frontal interactions, personal space, and defensive behavior. *Neuropsychologia*, 44(13):2621–2635.

Green, J. J. and McDonald, J. J. (2006). An event-related potential study of supramodal attentional control and crossmodal attention effects. *Psychophysiology*, 43(2):161–171.

Green, J. J., Teder-Sälejärvi, W. A., and McDonald, J. J. (2005). Control mechanisms mediating shifts of attention in auditory and visual space: A spatio-temporal ERP analysis. *Experimental Brain Research*, 166(3-4):358–369.

Green, P., Levison, W., Paelke, G., and Serafin, C. (1995). Preliminary human factors design guidelines for driver information systems. Technical report, US Department of Transportation, Federal Highway Administration.

Griffiths, T., Rees, G., Rees, A., Green, G. G. R., Witton, C., Rowe, D., Büchel, C., Turner, R., and Frackowiak, R. S. J. (1998). Right parietal cortex is involved in the perception of sound movement in humans. *Nat Neurosci*, 1(1):74–79.

Griffiths, T. D., Green, G. G. R., Rees, A., and Rees, G. (2000). Human brain areas involved in the analysis of auditory movement. *Human Brain Mapping*, 9:72–80.

Groppe, D. M., Urbach, T. P., and Kutas, M. (2011). Mass univariate analysis of event-related brain potentials/fields I: A critical tutorial review. *Psychophysiology*, 48(12):1711–25.

Guski, R. (1992). Acoustic tau: An easy analogue to visual tau?

Haas, E. C. and Edworthy, J. (1996). Designing urgency into auditory warnings using pitch, speed and loudness. *Computing & Control Engineering Journal*, (1):193–198.

Haas, E. C. and Schmidt, J. (1995). Auditory icons as warning and advisory signals in the U.S. army battlefield combat identification system (BCIS). In *Proceedings of the Human Factors and Ergonomics Society Annual Meeting 39th Annual Meeting*, pages 999–1003.

Hakkinen, M. T. and Williges, B. H. (1984). Synthesized warning messages: Effects of an alerting cue in single- and multiple-function voice synthesis systems. *Human Factors*, 26(2):185–195.

Hansen, J. C., Dickstein, P. W., Berka, C., and Hillyard, S. A. (1983). Event-related potentials during selective attention to speech sounds. *Biological Psychology*, 16(3-4):211–224.

Harter, M. R., Miller, S. L., Price, N. J., LaLonde, M. E., and Keyes, A. L. (1989). Neural processes involved in directing attention. *Journal of Cognitive Neuroscience*, 1(3):223–237.

Haufe, S., Kim, J. W., Kim, I. H., Sonnleitner, A., Schrauf, M., Curio, G., and Blankertz, B. (2014). Electrophysiology-based detection of emergency braking intention in real-world driving. *Journal of Neural Engineering*, 11(5).

Hein, E., Rolke, B., and Ulrich, R. (2006). Visual attention and temporal discrimination: Differential effects of automatic and voluntary cueing. *Visual Cognition*, 13(1):29–50.

Henderson, J. M. and Macquistan, A. D. (1993). The spatial distribution of attention following an exogenous cue. *Perception & psychophysics*, 53(2):221–230.

Hillyard, S. A., Hink, R. F., Schwent, V. L., and Picton, T. W. (1973). Electrical signs of selective attention in the human brain. *Science*, 182(4108):177–180.

Hillyard, S. A. and Münte, T. F. (1984). Selective attention to color and location: An analysis with event-related brain potentials. *Perception and Psychophysics*, 36(2):185–98.

Hirsh, I. J., Reynolds, E. G., and Joseph, M. (1954). Intelligibility of different speech materials. *The Journal of the Acoustical Society of America*, 26(4):530–538.

Ho, C., Gray, R., and Spence, C. (2013). Role of audiovisual synchrony in driving head orienting responses. *Experimental Brain Research*, 227(4):467–76.

Ho, C. and Spence, C. (2005). Assessing the effectiveness of various auditory cues in capturing a driver's visual attention. *Journal of Experimental Psychology: Applied*, 11(3):157–174.

Ho, C. and Spence, C. (2006). Verbal interface design: Do verbal directional cues automatically orient visual spatial attention? *Computers in Human Behavior*, 22(4):733–748.

Hopfinger, J. B., Buonocore, M. H., and Mangun, G. R. (2000). The neural mechanisms of top- down attentional control. *Nature Neuroscience*, 3(3):284–291.

Hunt, A. R. and Kingstone, A. (2003). Covert and overt voluntary attention: Linked or independent? *Cognitive Brain Research*, 18(1):102–105.

James, W. (1890). *The Principles of Psychology*. Henry Holt, New York.

Janssen, I., Heymsfield, S. B., Wang, Z., Ross, R., Kung, T. A., Cederna, P. S., Meulen, J. H. V. D., Urbanchek, M. G., Kuzon, M., Faulkner, J. A., Biol, J. G. A., Med, S., Morton, D., Rankin, P., Kent, L., and Dysinger, W. (2010). Skeletal muscle mass and distribution in 468 men and women aged 18 – 88 yr. *Journal Applied Physiology*, 89:81–88.

JASP Team (2018). JASP (Version 0.8.5) [Computer software].

Jonides, J. (1981). Voluntary versus automatic control over the mind's eye's movement. In *Attention and Performance IX*, pages 187–203.

Jung, T.-P., Makeig, S., Humphries, C., Lee, T.-W., McKeown, M. J., Iragui, V., and Sejnowski, T. J. (2000). Removing electroencephalographic artifacts by blind source separation. *Psychophysiology*, 37:163–178.

Kahneman, D. (1973). Attention and effort. *The American Journal of Psychology*, 88(2):339.

Kantowitz, B. H. and Knight, J. L. (1976). Testing tapping time-sharing. *Journal of Experimental Psychology*, 103(2):331–336.

Kastner, S., Pinsk, M. A., De Weerd, P., Desimone, R., and Ungerleider, L. G. (1999). Increased activity in human visual cortex during directed attention in the absence of visual stimulation. *Neuron*, 22(4):751–761.

Kazem, M. L. N., Noyes, J. M., and Lieven, N. J. (2003). Design considerations for a background auditory display to aid pilot situation awareness. In *Proceedings of the 2003 International Conference on Auditory Display*, pages 91–94.

Keitel, C., Maess, B., Schröger, E., and Müller, M. M. (2013). Early visual and auditory processing rely on modality-specific attentional resources. *NeuroImage*, 70:240–249.

Keller, P. and Stevens, C. (2004). Meaning from environmental sounds: Types of signal-referent relations and their effect on recognizing auditory icons. *Journal of Experimental Psychology: Applied*, 10(1):3–12.

Kiesel, A., Steinhauser, M., Wendt, M., Falkenstein, M., Jost, K., Philipp, A. M., and Koch, I. (2010). Control and interference in task switching – a review. *Psychological Bulletin*, 136(5):849–74.

King, S. M., Dykeman, C., Redgrave, P., and Dean, P. (1992). Use of a distracting task to obtain defensive head movements to looming visual stimuli by human adults in a laboratory setting. *Perception*, 21(2):245–259.

Klein, R. M. (2000). Inhibition of return. *Trends in Cognitive Sciences*, 4(4):138–147.

Kleiner, M., Brainard, D., Pelli, D., Ingling, A., Murray, R., and Broussard, C. (2007). What ' s new in Psychtoolbox-3 ? *Perception*, 36(14):1.

Knight, R. T. (1984). Decreased response to novel stimuli after prefrontal lesions in man. *Electroencephalography and clinical neurophysiology*, 59:9–20.

Koelewijn, T. (2009). *Audiovisual Attention in Space*. PhD thesis.

Kornhuber, H. and Deecke, L. (1964). Hirnpotentialänderungen beim Menschen vor und nach Willkürbewegungen dargestellt mit Magnetbandspeicherung und Rückwärtsanalyse. *Pflugers Archiv-European Journal of Physiology*, 281(1):52.

Kotz, S. A. (2014). Electrophysiological indices of speech processing. In *Encyclopedia of Computational Neuroscience*, pages 1–5.

Kraus, N. and Nicol, T. (2009). Auditory evoked potentials. In *Encyclopedia of Neuroscience*, pages 154–159.

Krupenia, S., Selmarker, A., Fagerlönn, J., Delsign, K., Jansson, A., Sandblad, B., and Grane, C. (2014). The 'Methods for Designing Future Autonomous Systems' (MODAS) project: Developing the cab for a highly autonomous truck. *Advances in Human Aspects of Transportation: Part II*, pages 70–73.

Kun, A. L., Boll, S., and Schmidt, A. (2016). Shifting gears: User interfaces in the age of autonomous driving. *IEEE Pervasive Computing*, 15(1):32–38.

Kutas, M. and Hillyard, S. A. (1989). An electrophysiological probe of incidental semantic association. *Journal of cognitive neuroscience*, 1(1):38–49.

Le, T. H., Pardo, J. V., and Hu, X. (1998). 4 T-fMRI study of nonspatial shifting of selective attention: Cerebellar and parietal contributions. *Journal of neurophysiology*, 79(3):1535–1548.

Lee, D. N. (1976). A theory of visual control of braking based on information about time-to-collision. *Perception*, 5(4):437–459.

Lee, D. N. (2009). Lee ' s 1976 paper. *Perception*, 38:837–859.

Lee, J. and Spence, C. (2015). Audiovisual crossmodal cuing effects in front and rear space. *Frontiers in Psychology*, 6:1–10.

Lee, Y.-C., Lin, W.-C., King, J.-T., Ko, L.-W., Huang, Y.-T., and Cherng, F.-Y. (2014). An EEG-based approach for evaluating audio notifications under ambient sounds. *Proceedings of the 32nd annual ACM conference on Human factors in computing systems - CHI '14*, pages 3817–3826.

Leech, R. and Sharp, D. J. (2014). The role of the posterior cingulate cortex in cognition and disease. *Brain*, 137(1):12–32.

Leo, F., Romei, V., Freeman, E., Ladavas, E., and Driver, J. (2011). Looming sounds enhance orientation sensitivity for visual stimuli on the same side as such sounds. *Experimental Brain Research*, 213(2-3):193–201.

Leung, Y. K., Smith, S., Parker, S., and Martin, R. (1997). Learning and retention of auditory warnings. *Proceedings of the Third International Conference on Auditory Display*.

Levitt, H. (1971). Transformed up-down methods in psychoacoustics. *The Journal of the Acoustical Society of America*, 49(2B):467–477.

Libet, B. (1985). Unconscious cerebral initiative and the role of conscious will in voluntary action. *Behavioral and Brain Sciences*, 8(4):529–539.

Liljedahl, M. and Fagerlönn, J. (2010). Methods for sound design: A review and implications for research and practice. In *Proceedings of the 5th Audio Mostly Conference: A Conference on Interaction with Sound*, pages 1–8.

Linden, D. E., Prvulovic, D., Formisano, E., Völlinger, M., Zanella, F. E., Goebel, R., and Dierks, T. (1999). The functional neuroanatomy of target detection: An fMRI study of visual and auditory oddball tasks. *Cerebral cortex*, 9(8):815–823.

Löcken, A., Borojeni, S. S., Müller, H., Gable, T. M., Triberti, S., Diels, C., Glatz, C., Alvarez, I., Chuang, L., and Boll, S. (2017). Towards adaptive ambient in-vehicle displays and interactions: Insights and design guidelines from the 2015 AutomotiveUI dedicated workshop. In *Automotive User Interfaces*, pages 325–348.

Löcken, A., Heuten, W., and Boll, S. (2015). Supporting lane change decisions with ambient light. In *Automotive User Interfaces '15*, pages 204–211.

Lopez-Calderon, J. and Luck, S. J. (2014). ERPLAB: An open-source toolbox for the analysis of event-related potentials. *Frontiers in human neuroscience*, 8(213):1–14.

Lu, S. A., Wickens, C. D., Prinet, J. C., Hutchins, S. D., Sarter, N., and Sebok, A. (2013). Supporting interruption management and multimodal interface design: Three meta-analyses of task performance as a function of interrupting task modality. *Human Factors: The Journal of the Human Factors and Ergonomics Society*, 55(4):697–724.

Lu, T., Liang, L., Wang, X., Mitchell, J. F., Reynolds, J. H., Miller, C. T., and Liang, L. I. (2001). Neural representations of temporally asymmetric stimuli in the auditory cortex of awake primates. *Journal of Neurophysiology*, 85:2364–2380.

Lu, Z. L., Tse, H. C. H., Dosher, B. A., Lesmes, L. A., Posner, C., and Chu, W. (2009). Intra- and cross-modal cuing of spatial attention: Time courses and mechanisms. *Vision Research*, 49(10):1081–1096.

Lucas, P. A. (1994). An evaluation of the communicative ability of auditory icons and earcons. In *Accepted for publication in the proceedings of ICAD*. Georgia Institute of Technology.

Luck, S. J. (2005). *An introduction to the event-relatedpotential technique*. A Bradford Book.

Ludewig, K., Ludewig, S., Seitz, A., Obrist, M., Geyer, M. A., and Vollenweider, F. X. (2003). The acoustic startle reflex and its modulation: Effects of age and gender in humans. *Biological Psychology*, 63(3):311–323.

Lupianez, J., Klein, R. M., and Bartolomeo, P. (2006). Inhibition of return: Twenty years after. *Cognitive neuropsychology*, 23(7):1003–14.

Macaluso, E. (2010). Orienting of spatial attention and the interplay between the senses. *Cortex*, 46(3):282–297.

Macmillan, N. A. and Creelman, C. D. (1991). *Detection theory: A user's guide*. Lawrence Erlbaum Associates, Inc Hillsdale, NJ, New Yersey, 1st edition.

Maddock, R. J., Garrett, A. S., and Buonocore, M. H. (2003). Posterior cingulate cortex activation by emotional words: fMRI evidence from a valence decision task. *Human Brain Mapping*, 18(1):30–41.

Maeder, P. P., Meuli, R. A., Adriani, M., Bellmann, A., Fornari, E., Thiran, J. P., Pittet, A., and Clarke, S. (2001). Distinct pathways involved in sound recognition and localization: A human fMRI study. *NeuroImage*, 14(4):802–816.

Maier, J. X., Chandrasekaran, C., and Ghazanfar, A. A. (2008). Integration of bimodal looming signals through neuronal coherence in the temporal lobe. *Current Biology*, 18(13):963–8.

Maier, J. X. and Ghazanfar, A. A. (2007). Looming biases in monkey auditory cortex. *The Journal of Neuroscience*, 27(15):4093–4100.

Maier, J. X., Neuhoff, J. G., Logothetis, N. K., and Ghazanfar, A. A. (2004). Multisensory integration of looming signals by rhesus monkeys. *Neuron*, 43:177–181.

Mangun, G. R. (1994). Orienting attention in the visual fields: An electrophysiological analysis. In Heinze, H.-J., Münte, T. F., and Mangun, G. R., editors, *Cognitive Electrophysiology*, pages 81–101. Birkhäuser, Boston, MA.

Mangun, G. R. and Hillyard, S. A. (1991). Modulation of sensory-evoked brain potentials provide evidence for changes in perceptual processing during visual-spatial priming. *Journal of Experimental Psychology: Human Perception and Performance*, 17(4):1057–1074.

Marshall, D. C., Lee, J. D., and Austria, P. A. (2007). Alerts for in-vehicle information systems: Annoyance, urgency, and appropriateness. *Human Factors: The Journal of the Human Factors and Ergonomics Society*, 49(1):145–157.

McCarthy, L. and Olsen, K. N. (2017). A looming bias in spatial hearing? Effects of acoustic intensity and spectrum on categorical sound source localization. *Attention, Perception, & Psychophysics*, 79(1):352–362.

McCrickard, D. S. and Chewar, C. M. (2003). Attuning notification design to user goals and attention costs. *Communications of the ACM*, 46(3):67.

McDonald, J. J., Teder-Sälejärvi, W. A., and Hillyard, S. A. (2000). Involuntary orienting to sound improves visual perception. *Nature*, 407(6806):906–908.

McDonald, J. J., Teder-Sälejärvi, W. A., and Ward, L. M. (2001). Multisensory integration and crossmodal attention effects in the human brain. *Science*,

292(5523):1791.

McGookin, D. and Brewster, S. (2004). Understanding concurrent earcons: Applying auditory scene analysis principles to concurrent earcon recognition. *ACM Transactions on Applied Perception (TAP)*, 1(2):130–155.

McGuire, A. B., Gillath, O., and Vitevitch, M. S. (2016). Effects of mental resource availability on looming task performance. *Attention, Perception, & Psychophysics*, 78(1):107–113.

McKeown, D. (2005). Candidates for within-vehicle auditory displays. In *Proceedings of ICAD 05 - Eleventh Meeting of the International Conference on Auditory Display*, pages 182–189.

McKeown, D. and Isherwood, S. (2007). Mapping candidate within-vehicle auditory displays to their referents. *Human Factors: The Journal of the Human Factors and Ergonomics Society*, 49(3):417–428.

McLeod, P. (1977). A dual task response modality effect: Support for multiprocessor models of attention. *Quarterly Journal of Experimental Psychology*, 29(4):651–667.

Meredith, M. A., Nemitz, J. W., and Stein, B. E. (1987). Determinants of multisensory integration in superior colliculus neurons. I. Temporal factors. *The Journal of neuroscience*, 7(10):3215–29.

Mesulam, M. (1999). Spatial attention and neglect: Parietal, frontal and cingulate contributions to the mental representation and attentional targeting of salient extrapersonal events. *Philosophical Transactions of the Royal Society B: Biological Sciences*, 354(1392):1325–1346.

Middlebrooks, J. C. and Green, D. M. (1991). Sound localization by human listeners. *Annual Review of Psychology*, 42(1):135–159.

Miller, E. K. and Cohen, J. D. (2001). An integrative theory of prefrontal cortex function. *Annu Rev Neurosci*, 24:167–202.

Mohanty, A., Gitelman, D. R., Small, D. M., and Mesulam, M. M. (2008). The spatial attention network interacts with limbic and monoaminergic systems to modulate motivation-induced attention shifts. *Cerebral Cortex*, 18(11):2604–2613.

Molfese, D. L. (1983). Event related potentials and language processes. In Gaillard, A. W. K. and Ritter, W., editors, *Tutorials in ERP Research: Endogenous Components*, pages 345–368. North-Holland Publishing Company.

Mollenhauer, M. A., Lee, J., Cho, K., Hulse, M. C., and Dingus, T. A. (1994). The effects of sensory modality and information priority on in-vehicle signing and information systems. In *Proceedings of the Human Factors and Ergonomics Society Annual Meeting*, volume 38, pages 1072–1076.

Mondor, T. A. and Zatorre, R. J. (1995). Shifting and focusing auditory spatial attention. *Journal of Experimental Psychology: Human Perception and Performance*, 21(2):387–409.

Monsell, S. (1996). Control of mental processes. In *Unsolved Mysteries of the Mind: Tutorial Essays in Cognition*, pages 93–148. Erlbaum.

Monsell, S. (2003). Task switching. *Trends in Cognitive Sciences*, 7(3):134–140.

Moray, N. (1967). Where is capacity limited? A survey and a model. *Acta Psychologica*, 27:84–92.

Müller, H. J. and Rabbitt, P. M. (1989). Reflexive and voluntary orienting of visual attention: Time course of activation and resistance to interruption. *Journal of Experimental Psychology: Human Perception and Performance*, 15(2):315–30.

Mynatt, E. D. (1994). Designing with auditory icons: How well do we identify auditory cues? In *Conference on Human Factors in Computing Systems (CHI'94)*, pages 269–270.

Näätänen, R., Gaillard, A. W. K., and Mäntysalo, S. (1978). Early selective-attention effect on evoked potential reinterpreted. *Acta Psychologica*, 42(4):313–329.

Näätänen, R., Muranen, V., and Merisalo, A. (1974). Timing of expectancy peak in simple reaction time situation. *Acta Psychologica*, 38(6):461–470.

Näätänen, R. and Picton, T. W. (1987). The N1 wave of the human electric and magnetic response to sound: A review and an analysis of the component structure. *Psychophysiology*, 24(4):375–425.

Nakayama, K. and Mackeben, M. (1989). Sustained and transient components of focal visual attention. *Vision research*, 29(11):1631–1647.

Navon, D. and Gopher, D. (1979). On the economy of the human-processing system. *Psychological Review*, 86(3):214–255.

Nees, M. A. and Walker, B. N. (2005). Auditory interfaces and sonification. In *The Universal Access Handbook*, pages 507–522.

Nees, M. A. and Walker, B. N. (2011). Auditory displays for in-vehicle technologies. *Reviews of Human Factors and Ergonomics*, 7(1):58–99.

Neuhoff, J. G. (1998). Perceptual bias for rising tones. *Nature*, 395(6698):123–4.

Neuhoff, J. G. (2001). An adaptive bias in the perception of looming auditory motion. *Ecological Psychology*, 13(2):87–110.

Neuhoff, J. G. (2016). Looming sounds are perceived as faster than receding sounds. *Cognitive Research: Principles and Implications*, pages 1–9.

Neuhoff, J. G., Bilecen, D., Mustovic, H., Schachinger, H., Seifritz, E., Scheffler, K., and Di Salle, F. (2002). A cortical network underpinning the perceptual priority for rising intensity and auditory "looming". *The Journal of the Acoustical Society of America*, 111(5):2355–2355.

Neuhoff, J. G., Long, K. L., and Worthington, R. C. (2012). Strength and physical fitness predict the perception of looming sounds. *Evolution and Human Behavior*, 33(4):318–322.

Neuhoff, J. G., Planisek, R., and Seifritz, E. (2009). Adaptive sex differences in auditory motion perception: Looming sounds are special. *Journal of Experimental Psychology: Human Perception and Performance*, 35(1):225–234.

Nieuwenhuizen, F. M., Mulder, M., van Paassen, M. M., and Bülthoff, H. H. (2013). Influences of simulator motion system characteristics on pilot control behavior. *Journal of Guidance, Control, and Dynamics*, 36(3):667–676.

Nobre, A. C., Gitelman, D. R., Dias, E. C., and Mesulam, M. M. (2000a). Covert visual spatial orienting and saccades: Overlapping neural systems. *NeuroImage*, 11(3):210–216.

Nobre, A. C., Sebestyen, G. N., and Miniussi, C. (2000b). The dynamics of shifting visuospatial attention revealed by event-related potentials. *Neuropsychologia*, 38(7):964–974.

Noel, J. P., Grivaz, P., Marmaroli, P., Lissek, H., Blanke, O., and Serino, A. (2015). Full body action remapping of peripersonal space: The case of walking. *Neuropsychologia*, 70:375–384.

Novak, G., Ritter, W., and Vaughan, H. G. (1992). Mismatch detection and the latency of temporal judgements. *Psychophysiology*, 29(4):398–411.

Noyes, J. M., Hellier, E., and Edworthy, J. (2006). Speech warnings: A review. *Theoretical Issues in Ergonomics Science*, 7(6):551–571.

Nykänen, A. (2008). *Methods for Product Sound Design*. PhD thesis.

O'Callaghan, C. (2009). Auditory perception.

Oh, E. L. and Lutfi, R. A. (1999). Informational masking by everyday sounds. *The Journal of the Acoustical Society of America*, 106(6):3521–3528.

Olsen, K. N. (2014). Intensity dynamics and loudness change: A review of methods and perceptual processes. *Acoustics Australia*, 42(3):159–165.

Olsen, K. N. and Stevens, C. J. (2010). Perceptual overestimation of rising intensity: Is stimulus continuity necessary? *Perception*, 39(5):695–704.

Orgs, G., Lange, K., Dombrowski, J. H., and Heil, M. (2006). Conceptual priming for environmental sounds and words: An ERP study. *Brain and Cognition*, 62(3):267–272.

Pashler, H. (1998). *The psychology of attention*. The MIT press.

Pashler, H. (2000). Task switching and multitask performance. In Monsell, S. and Driver, J., editors, *Control of cognitive processes: Attention and performance XVIII*, pages 275–307.

Paul, C. L., Komlodi, A., and Lutters, W. (2015). Interruptive notifications in support of task management. *International Journal of Human Computer Studies*, 79:20–34.

Pelli, D. G. (1997). The VideoToolbox software for visual psychophysics: Transforming numbers into movies. *Spatial Vision*, 10(4):437–442.

Petersen, S. E. and Posner, M. I. (2012). The attention system of the human brain: 20 years after. *Annual review of neuroscience*, 21(35):73–89.

Petersen, S. E., Robinson, D. L., and Currie, J. N. (1989). Influences of lesions of parietal cortex on visual spatial attention in humans. *Experimental Brain Research*, 76(2):267–80.

Piazza, C., Miyakoshi, M., Akalin-Acar, Z., Cantiani, C., Reni, G., Bianchi, A. M., and Makeig, S. (2016). An automated function for identifying EEG independent components representing bilateral source activity. *XIV Mediterranean Conference on Medical and Biological Engineering and Computing 2016, IFMBE Proceedings*, 57:105–109.

Picton, T. W. (1992). The P300 wave of the human event-related potential. *Journal of clinical neurophysiology: official publication of the American Electroencephalographic Society*, 9(4):456–479.

Picton, T. W. (2011). *Human auditory evoked potentials*. Plural Publishing.

Picton, T. W. (2014). Auditory event-related potentials. *Encyclopedia of Computational Neuroscience*, pages 1–6.

Picton, T. W. and Hillyard, S. A. (1974). Human auditory evoked potentials. II: Effects of attention. *Electroencephalography and Clinical Neurophysiology*, 36:191–199.

Picton, T. W., Hillyard, S. A., and Galambos, R. (1976). Habituation and attention in the auditory system. In *Auditory System*, pages 343–389.

Polich, J. (2003). Theoretical overview of P3a and P3b. In *Detection of Change: Event-related potential and fMRI findings*, pages 83 – 98.

Polich, J. (2007). Updating P300: An integrative theory of P3a and P3b. *Clinical Neurophysiology*, 118:2128–2148.

Politis, I., Brewster, S., and Pollick, F. (2015). Language-based multimodal displays for the handover of control in autonomous cars. *Proceedings of the 7th International Conference on Automotive User Interfaces and Interactive Vehicular Applications - AutomotiveUI '15*, (c):3–10.

Popper, A. N. and Fay, R. R. (1997). Evolution of the ear and hearing: Issues and questions. *Brain Behav Evol*, 50(4):213–221.

Posner, M. I. (1978). *Chronometric explorations of mind*. Lawrence Erlbaum Associates, Inc Hillsdale, NJ.

Posner, M. I. (1980). Orienting of attention. *The Quarterly Journal of Experimental Psychology*, 32(1):3–25.

Posner, M. I. and Boies, S. J. (1971). Components of attention. *Psychological Review*, 78(5):391–408.

Posner, M. I. and Cohen, Y. (1984). Components of visual orienting. In *Attention and Performance: Control of Language Processes*, chapter 32, pages 531–556.

Posner, M. I. and Petersen, S. E. (1990). The attention system of the human brain. *Annual Review of Neuroscience*, 13:25–42.

Posner, M. I., Rafal, R. D., Choate, L. S., and Vaughan, J. (1985). Inhibition of return: Neural basis and function. *Cognitive Neuropsychology*, 2(3):211–228.

Posner, M. I., Snyder, C. R. R., and Davidson, B. J. (1980). Attention and the detection of signals. *Journal of Experimental Psychology: General*, 109(2):160–174.

Pousman, Z. and Stasko, J. (2006). A taxonomy of ambient information systems. *Proceedings of the working conference on Advanced visual interfaces - AVI '06*, page 67.

Praamstra, P., Stegeman, D., Horstink, M., and Cools, A. (1996). Dipole source analysis suggests selective modulation of the supplementary motor area contribution to the readiness potential. *Electroencephalography and Clinical Neurophysiology*, 98(6):468–477.

Pulvermüller, F. (1999). Words in the brain's language. *Behavioral and Brain Sciences*, 22:253–336.

Rapin, I., Schimmel, H., Tourk, L. M., Krasnegor, N. A., and Pollak, C. (1966). Evoked responses to clicks and tones of varying intensity in waking adults. *Electroencephalography and Clinical Neurophysiology*, 21(4):335–344.

Ratcliff, R. (1993). Methods for dealing with reaction-time outliers. *Psychological Bulletin*, 114(3):510–532.

Riggins, B. R. and Polich, J. (2002). Habituation of P3a and P3b from visual stimuli. *The Korean Journal of Thinking and Problem Solving*, 12(1):71–81.

Riskind, J. H., Kleiman, E. M., Seifritz, E., and Neuhoff, J. G. (2014). Influence of anxiety, depression and looming cognitive style on auditory looming perception. *Journal of Anxiety Disorders*, 28(1):45–50.

Ro, T., Farnè, A., and Chang, E. (2003). Inhibition of return and the human frontal eye fields. *Experimental Brain Research*, 150(3):290–296.

Romei, V., Murray, M. M., Cappe, C., and Thut, G. (2009). Preperceptual and stimulus-selective enhancement of low-level human visual cortex excitability by sounds. *Current Biology*, 19(21):1799–805.

Rosenblum, L. D., Carello, C., and Pastore, R. E. (1987). Relative effectiveness of three stimulus variables for locating a moving sound source. *Perception*, 16(2):175–186.

Rosenblum, L. D., Wuestefeld, A. P., and Saldaña, H. M. (1993). Auditory looming perception: Influences on anticipatory judgments. *Perception*, 22(12):1467–1482.

Rosenholtz, R. (2011). What your visual system sees where you are not looking. *Proc SPIE Human Vision and Electronic Imaging*, 7(1):786510–786510–14.

Ross, T., Midtland, K., Fuchs, M., Pauzié, A., Engert, A., Duncan, B., Vaughan, G., Vernet, M., Peters, H., Burnett, G., and May, A. (1996). HARDIE design guidelines handbook. Technical report.

Salces, F. J. S. and Vickers, P. (2003). Household appliances control device for the elderly. In *Proceedings of the 2003 International Conference on Auditory Display*, pages 224–227.

Santangelo, V. and Spence, C. (2007). Multisensory cues capture spatial attention regardless of perceptual load. *Journal of experimental psychology. Human perception and performance*, 33(6):1311–21.

Saygin, A. P., Dick, F., and Bates, E. (2005). An on-line task for contrasting auditory processing in the verbal and nonverbal domains and norms for younger and older adults. *Behavior research methods*, 37(1):99–110.

Saygin, A. P., Dick, F., Wilson, S. W., Dronkers, N. F., and Bates, E. (2003). Neural resources for processing language and environmental sounds: Evidence from aphasia. *Brain*, 126(4):928–945.

Scheer, M. (2017). *Auditory distraction during visuomotor steering*. PhD thesis.

Scheer, M., Bülthoff, H. H., and Chuang, L. L. (2016). Steering demands diminish the early-P3, late-P3 and RON components of the event-related potential of task-irrelevant environmental sounds. *Frontiers in Human Neuroscience*, 10(73):1–15.

Schiff, W., Caviness, J. A., and Gibson, J. J. (1962). Persistent fear responses in rhesus monkeys to the optical stimulus of "looming". *Science*, 136:982–983.

Schiff, W. and Oldak, R. (1990). Accuracy of judging time to arrival: Effects of modality, trajectory, and gender. *Journal of Experimental Psychology: Human Perception and Performance*, 16(2):303–316.

Schmitt, M., Postma, A., and De Haan, E. (2000). Interactions between exogenous auditory and visual spatial attention. *Quarterly Journal of Experimental Psychology Section A-Human Experimental Psychology*, 53(1):105–130.

Schomer, D. L. and Da Silva, F. L. (2012). *Niedermeyer's electroencephalography: Basic principles, clinical applications, and related fields*. Lippincott Williams and Wilkins.

Schouten, B., Troje, N. F., Vroomen, J., and Verfaillie, K. (2011). The effect of looming and receding sounds on the perceived in-depth orientation of depth-ambiguous biological motion figures. *PLoS ONE,* 6(2):1–8.

Schröger, E. (1993). Event-related potentials to auditory stimuli following transient shifts of spatial attention in a Go/Nogo task. *Biological Psychology,* 36(3):183–207.

Schröger, E. (1994). Human brain potential signs of selection by location and frequency in an auditory transient attention situation. *Neuroscience Letters,* 173(1-2):163–166.

Schröger, E. and Eimer, M. (1997). Endogenous covert spatial orienting in audition: Cost-benefit analyses of reaction times and event-related potentials. *The Quarterly Journal of Experimental Psychology,* 50A(2):457–474.

Schröger, E., Marzecová, A., and Sanmiguel, I. (2015). Attention and prediction in human audition: A lesson from cognitive psychophysiology. *European Journal of Neuroscience,* 41(5):641–664.

Schultze-Kraft, M., Birman, D., Rusconi, M., Allefeld, C., Görgen, K., Dähne, S., Blankertz, B., and Haynes, J.-D. (2016). The point of no return in vetoing self-initiated movements. *Proceedings of the National Academy of Sciences,* 113(4):1080–1085.

Seifritz, E., Neuhoff, J. G., Bilecen, D., Scheffler, K., Mustovic, H., Schächinger, H., Elefante, R., and Di Salle, F. (2002). Neural processing of auditory looming in the human brain. *Current Biology,* 12(24):2147–51.

Selcon, S. J., Taylor, R. M., and McKenna, F. P. (1995). Integrating multiple information sources: Using redundancy in the design of warnings. *Ergonomics,* 38(11):2362–2370.

Sestieri, C., Di Matteo, R., Ferretti, A., Del Gratta, C., Caulo, M., Tartaro, A., Olivetti Belardinelli, M., and Romani, G. L. (2006). "What" versus "Where" in the audiovisual domain: An fMRI study. *NeuroImage,* 33(2):672–680.

Shahin, A., Bosnyak, D. J., Trainor, L. J., and Roberts, L. E. (2003). Enhancement of neuroplastic P2 and N1c auditory evoked potentials in musicians. *Journal of Neuroscience,* 23(13):5545–5552.

Shaw, B. K., McGowan, R. S., and Turvey, M. T. (1991). An acoustic variable specifying time-to-contact. *Ecolgocial Psychology,* 3(3):253–261.

Shibasaki, H. and Hallett, M. (2006). What is the Bereitschaftspotential? *Clinical Neurophysiology,* 117(11):2341–2356.

Simon, J. R. and Craft, J. L. (1970). Effects of an irrelevant auditory stimulus on visual choice reaction time. *Journal of Experimental Psychology,* 86(2):272–274.

Simon, O., Mangin, J. F., Cohen, L., Le Bihan, D., and Dehaene, S. (2002). Topographical layout of hand, eye, calculation, and language-related areas in the human parietal lobe. *Neuron,* 33(3):475–487.

Simpson, C. A. (1976). Effects of linguistic redundancy on pilot's comprehension of synthesized speech. In *Twelfth Annual Conference on Manual Control,* pages 294–308.

Simpson, C. A. and Marchionda-Frost, K. (1984). Synthesized speech rate and pitch effects on intelligibility of warning messages for pilots. *Human factors,* 26(5):509–517.

Simpson, C. A. and Williams, D. H. (1980). Response time effects of alerting tone and semantic context for synthesized voice cockpit warnings. *Human Factors:*

The Journal of the Human Factors and Ergonomics Society, 22(3):319–330.

Skarratt, P. A., Cole, G. G., and Gellatly, A. R. H. (2009). Prioritization of looming and receding objects: Equal slopes, different intercepts. *Attention, Perception, and Psychophysics*, 71(4):964–970.

Snyder, E. and Hillyard, S. A. (1976). Long-latency evoked potentials to irrelevant, deviant stimuli. *Behavioral Biology*, 16(3):319–331.

Spence, C. (2010). Crossmodal spatial attention. *Annals of the New York Academy of Sciences*, 1191:182–200.

Spence, C. and Driver, J. (1994). Covert spatial orienting in audition: Exogenous and endogenous mechanisms. *Journal of Experimental Psychology: Human Perception and Performance*, 20(3):555–574.

Spence, C. and Driver, J. (1996). Audiovisual links in endogenous covert spatial attention. *Journal of Experimental Psychology: Human Perception and Performance*, 22(4):1005–1030.

Spence, C. and Driver, J. (1997). Audiovisual links in exogenous covert spatial orienting. *Perception and Psychophysics*, 59(1):1–22.

Spence, C. and Driver, J. (1998). Auditory and audiovisual inhibition of return. *Perception and Psychophysics*, 60(1):125–139.

Spence, C. and Driver, J. (2004). *Crossmodal space and crossmodal attention*. Oxford University Press.

Spence, C., McDonald, J., and Driver, J. (2004). Exogenous spatial cuing studies of human crossmodal attention and multisensory integration. In *Crossmodal space and crossmodal attention*, pages 277–320.

Spence, C. and McGlone, F. P. (2001). Reflexive spatial orienting of tactile attention. *Experimental Brain Research*, 141(3):324–330.

Spence, C., Nicholls, M. E., Gillespie, N., and Driver, J. (1998). Cross-modal links in exogenous covert spatial orienting between touch, audition, and vision. *Perception and Psychophysics*, 60(4):544–557.

Spence, C. and Read, L. (2003). Speech shadowing while driving: On the difficulty of splitting attention between eye and ear. *Psychological Science*, 14(3):251–256.

Squires, N. K., Squires, K. C., and Hillyard, S. A. (1975). Two varieties of long-latency positive waves evoked by unpredictable auditory stimuli in man. *Electroencephalography and clinical neurophysiology*, 38:387–401.

Stein, B. E., London, N., Wilkinson, L. K., and Price, D. D. (1996). Enhancement of perceived visual intensity by auditory stimuli: A psychophysical analysis. *Journal of Cognitive Neuroscience*, 8(6):497–506.

Steinschneider, M. and Dunn, M. (2002). Electrophysiology in developmental neuropsychology. *Handbook of Neuropsychology*, 8(1):91–146.

Stevens, A., Quimby, A., Board, A., Kersloot, T., and Burns, P. (2002). Design guidelines for safety of in-vehicle information systems. *Department of Transport, Local Government and the Regions*, pages 1–55.

Strayer, D. L. and Johnston, W. A. (2001). Driven to distraction: Dual-task studies of simulated driving and conversing on a cellular telephone. *Psychological Science*, 12(6):462–466.

Sturm, W. and Willmes, K. (2001). On the functional neuroanatomy of intrinsic and phasic alertness. *NeuroImage*, 14(1 II):76–84.

Susini, P., McAdams, S., and Smith, B. K. (2002). Global and continuous loudness estimation of time-varying levels. *Acta Acustica united with Acustica*, 88:536–548.

Susini, P., Meunier, S., Trapeau, R., and Chatron, J. (2010). End level bias on direct loudness ratings of increasing sounds. *The Journal of the Acoustical Society of America*, 128(4):163–168.

Sutherland, C. A. M., Thut, G., and Romei, V. (2014). Hearing brighter: Changing in-depth visual perception through looming sounds. *Cognition*, 132(3):312–323.

Szekely, A., D'Amico, S., Devescovi, A., Federmeier, K., Herron, D., Iyer, G., Jacobsen, T., Anal'a, L. A., Vargha, A., and Bates, E. (2005). Timed action and object naming. *Cortex*, 41(1):7–25.

Talsma, D., Doty, T. J., Strowd, R., and Woldorff, M. G. (2006). Attentional capacity for processing concurrent stimuli is larger across sensory modalities than within a modality. *Psychophysiology*, 43(6):541–549.

Tassinari, G. and Campara, D. (1996). Consequences of covert orienting to non-informative stimuli of different modalities: A unitary mechanism? *Neuropsychologia*, 34(3):235–245.

Teder-Sälejärvi, W., Münte, T. F., Sperlich, F., and Hillyard, S. (1999). Intra-modal and cross-modal spatial attention to auditory and visual stimuli. An event-related brain potential study. *Cognitive brain research*, 8(3):327–343.

Teghtsoonian, R., Teghtsoonian, M., and Canévet, G. (2005). Sweep-induced acceleration in loudness change and the "bias for rising intensities". *Perception and Psychophysics*, 67(4):699–712.

Terry, H. R., Charlton, S. G., and Perrone, J. A. (2008). The role of looming and attention capture in drivers' braking responses. *Accident Analysis and Prevention*, 40(4):1375–1382.

Thiel, C. M., Zilles, K., and Fink, G. R. (2004). Cerebral correlates of alerting, orienting and reorienting of visuospatial attention: An event-related fMRI study. *NeuroImage*, 21(1):318–328.

Tinbergen, N. (1951). *The study of instinct*.

Tran, T. V., Letowski, T., and Abouchacra, K. S. (2000). Evaluation of acoustic beacon characteristics for navigation tasks. *Ergonomics*, 43(6):807–827.

Tremblay, K., Kraus, N., McGee, T., Ponton, C., and Otis, B. (2001). Central auditory plasticity: Changes in the N1-P2 complex after speech-sound training. *Ear and hearing*, 22(2):79–90.

Tremblay, K. L., Shahin, A. J., Picton, T., and Ross, B. (2009). Auditory training alters the physiological detection of stimulus-specific cues in humans. *Clinical Neurophysiology*, 120(1):128–135.

Tsimhoni, O. and Flannagan, M. J. (2006). Pedestrian detection with night vision systems enhanced by automatic warnings. In *Proceedings of the Human Factors and Ergonomics Society Annual Meeting*, number 50, pages 2443–2447.

Tuuri, K., Mustonen, M.-S., and Pirhonen, A. (2007). Same sound - different meanings: A novel scheme for modes of listening. *Audio Mostly 2007*, pages 13–18.

Tyll, S., Bonath, B., Schoenfeld, M. A., Heinze, H.-J., Ohl, F. W., and Noesselt, T. (2013). Neural basis of multisensory looming signals. *NeuroImage*, 65:13–22.

Ulfvengren, P. (2003). *Design of natural warning sounds in human-machine systems.* PhD thesis.

Van Coile, B., Rühl, H. W., Vogten, L., Thoone, M., Goß, S., Delaey, D., Moons, E., Terken, J. M., De Pijper, J. R., Kugler, M., Kaufholz, P., Krüger, R., Leys, S., and Willems, S. (1997). Speech synthesis for the new Pan-European traffic message control system RDS-TMC. *Speech Communication*, 23(4):307–317.

Van Petten, C. and Rheinfelder, H. (1995). Conceptual relationships between spoken words and environmenal sounds: Event-related brain potential measures. *Neuropsychologia*, 33(4):485–508.

Walker, B. N. and Nees, M. A. (2011). Theory of sonification. *The Sonification Handbook*, pages 9–39.

Ward, L. M. (1994). Supramodal and modality-specific mechanisms for stimulus-driven shifts of auditory and visual attention. *Canadian journal of experimental psychology*, 48(2):242–59.

Welford, A. T. (1952). The 'psychological refractory period' and the timing of high-speed performance - A review and a theroy. *British Journal of Psychology*, 43(4):2–19.

Welford, A. T. (1967). Single channel operation in the brain. *Acta psychologica*, 27:5–22.

Wickens, C., Kramer, A., Vanasse, L., and Donchin, E. (1983). Performance of concurrent tasks: A psychophysiological analysis of the reciprocity of information-processing resources. *Science*, 221(4615):1080–1082.

Wickens, C. D. (1976). The effects of divided attention on information processing in manual tracking. *Journal of Experimental Psychology: Human Perception and Performance*, 2(1):1–13.

Wickens, C. D. (1980). The structure of attentional resources. In *Attention and Performance VIII*, pages 239–258.

Wickens, C. D. (2002). Multiple resources and performance prediction. *Theoretical Issues in Ergonomics Science*, 3(2):159–177.

Wickens, C. D. (2008). Applied attention theory. *Ergonomics*, 1(2):222.

Wiese, E. E. and Lee, J. D. (2007). Attention grounding: A new approach to in-vehicle information system implementation. *Theoretical Issues in Ergonomics Science*, 8(3):255–276.

Winkler, I., Denham, S. L., and Escera, C. (2013). Auditory event-related potentials. In *Encyclopedia of Computational Neuroscience*, pages 1–29.

Woods, D. L. (1995). The component structrue of the N1 wave of the human auditory evoked potential. *Perspectives of Event-Related Potentials Research (Electroencephalography and Clinical Neurophysiology-Supplements Only)*, 44:102–109.

Yang, Z. and Mayer, A. R. (2014). An event-related fMRI study of exogenous orienting across vision and audition. *Human Brain Mapping*, 35(3):964–974.

Yantis, S. and Jonides, J. (1990). Abrupt visual onsets and selective attention: Voluntary versus automatic allocation. *Journal of Experimental Psychology: Human Perception and Performance*, 16(1):121–134.

Yeshurun, Y. and Carrasco, M. (1998). Attention improves or impairs visual performance by enhancing spatial resolution. *Nature*, 396(6706):1–8.

Yeshurun, Y. and Levy, L. (2003). Transient spatial attention degrades temporal resolution. *Psychological Science*, 14(3):225–231.

Yilmaz, M. and Meister, M. (2013). Rapid innate defensive responses of mice to looming visual stimuli. *Current Biology*, 23(20):2011–2015.

Zander, T. O., Andreessen, L. M., Berg, A., Bleuel, M., Pawlitzki, J., Zawallich, L., Krol, L. R., and Gramann, K. (2017). Evaluation of a dry EEG system for application of passive brain-computer interfaces in autonomous driving. *Frontiers in Human Neuroscience*, 11:1–16.

Zatorre, R. J., Bouffard, M., Ahad, P., and Belin, P. (2002). Where is 'where' in the human auditory cortex ? *Nature Neuroscience*, 5(9):905–909.

Zhang, S., Benenson, R., Omran, M., Hosang, J., and Schiele, B. (2016). How far are we from solving pedestrian detection? In *2016 IEEE Conference on Computer Vision and Pattern Recognition (CVPR)*, pages 1259–1267.